D0171008

Conceptualizing
Technological Change

Conceptualizing Technological Change

Theoretical and Empirical Explorations

Govindan Parayil

ROWMAN & LITTLEFIELD PUBLISHERS, INC.
Lanham • Boulder • New York • Oxford

ROWMAN & LITTLEFIELD PUBLISHERS, INC.

Published in the United States of America
by Rowman & Littlefield Publishers, Inc.
4720 Boston Way, Lanham, Maryland 20706
http://www.rowmanlittlefield.com

12 Hid's Copse Road
Cumnor Hill, Oxford OX2 9JJ, England

Copyright © 1999 by Rowman & Littlefield Publishers, Inc.

All rights reserved. No part of this publication may be reproduced,
stored in a retrieval system, or transmitted in any form or by any
means, electronic, mechanical, photocopying, recording, or otherwise,
without the prior permission of the publisher.

British Library Cataloguing in Publication Information Available

Library of Congress Cataloging-in-Publication Data
Parayil, Govindan, 1955-
    Conceptualizing technological change : theoretical and empirical
explorations / Govindan Parayil.
        p.   cm.
    Includes bibliographical references.
    ISBN 0-8476-9520-4 (cloth : alk. paper)
    1. Technology—Social aspects.   2. Technological innovations—
Social aspects.   I. Title.
T14.5.P37   1999
303.48'3—dc21   99-20491
                                                            CIP

Printed in the United States of America

♾ ™The paper used in this publication meets the minimum requirements of American
National Standard for Information Sciences—Permanence of Paper for Printed Library
Materials, ANSI Z39.48–1992.

Contents

Preface

The desire to study technology in society occurred to me more than fifteen years ago. That desire turned into an intellectual journey during the past several years, and consequently I developed a keen interest in investigating and explaining the seemingly inscrutable and abstract phenomenon of *technological change*. The more I read about technological change in any disciplinary field that has deliberated on this important phenomenon, the more I realized that none of the disciplines got it right when it came to explaining and theorizing technological change. Unfortunately, instead of enhancing my knowledge about this important phenomenon, whose centrality in our lives, for good or ill, is unprecedented, I became disheartened when it dawned on me that there was no consensus among the various disciplines (and authors) about what exactly constitutes technological change. In fact, we seem to know very little about how technology develops and changes. Therefore, one of the main objectives of this book is to fill this lacuna in the literature on explicating technological change.

Serious efforts have been made in this book to present a detailed theoretical and empirical account of technological change by drawing together current scholarship from disciplines ranging from history to sociology to economics and beyond. Yet, no claim is being made that the result is an encyclopedic account of this phenomenon. It would be rather presumptuous to make such a claim. However, I do believe that my findings in this book did break new theoretical and empirical grounds, such as explaining technological change as a problem-solving activity, and theorizing it as a reflexive social-historical process that is contingent upon the knowledge change of the relevant agents. I have shown that the best means to explain technology transfer is not by focusing on the tangible material artifacts involved in the transfer process, but rather by focusing on the exchange of the seemingly intangible ideas and information (knowledge) between human agents.

I wish to thank several colleagues, friends, and students who have given me valuable suggestions, advice, comments, and research assistance during the preparation of the manuscript. While some of them read and commented on the entire manuscript, others read selected chapters. And some colleagues commented on my initial outlines of the book project. Without mentioning the specific nature of their help or part in this project, I will just name them: Herbert Gottweis, Tom Misa, Steve Fuller, Joseph Pitt, Arif Dirlik, Robert Ferguson, Jesus Felipe, George Braine, Jennifer Johns, Greg Felker, Ho Wai-yip, and Cheung Chung Kit Carl. I should add a special word of thanks to Susan McEachern, my editor at Rowman & Littlefield, for her enthusiasm in getting the manuscript reviewed and putting it into the production process in a relatively short period of time. Without the encouragement and support of my wife, Olive, it would not have been possible for me to finish the manuscript in the time frame that I had set for it. A remarkable thing happened when the manuscript was completed, I became a father. Therefore, with great pleasure I dedicate this book to my wife, Olive, and son, Anik.

1

Introduction

TECHNOLOGICAL CHANGE: THE VERY IDEA

If there is one phenomenon that distinguishes the twentieth century from all the previous ones in human history, it has to be the spectacular changes that have been taking place in the technological knowledge base of modern and modernizing societies. No human enterprise can surpass technology's success in improving the material conditions, enhancing the cognitive attributes, and attenuating the physical limitations of humans. While some celebrate the coming of unprecedented wealth and prosperity brought on by the recent advances in genetic engineering, nanotechnology, computer-communication systems, and so on, others bemoan the possible demise of humans as autonomous social beings in the coming brave new world of cyborgs, transgenic organisms, and clones.[1] Yet, despite its centrality in our lives and cultures, for good or ill, we seem to know very little about technology and how it develops and changes both spatially and temporally. Despite the recognition of technological change as an important conceptual problem in science, technology, and society (STS) and its cognate domain of science and technology studies (S&TS), still, it is not understood well as a dynamic and multidimensional process that has important social and economic implications. When economists argue eloquently about the need to induce rapid technological change for productivity increase and economic growth, or when sociologists discourse discursively about technological change as a social process involving complex networks of actors contingently shaping things and society, are they implying the same thing and are they talking about the same phenomenon? What exactly is technological change? Why is it important to study technological change? How successful have been the efforts by the STS community and others, so far, to formally and informally theorize and model technological change? Can we apply symmetrically theories of scientific change to conceptualize technological change? What are the features that a comprehensive model of technological change should have? The objective of this book is to find possible answers to these and other related questions and concerns.

It has only been in the past two decades or so that efforts to explain, theorize, and model technological change became a serious research program in the field of science and technology studies. For example, a notable recent intellectual effort to focus explicitly on the methods and themes of technological change was attempted at the international conference on "Technological Change" held at Oxford University in 1993.[2] However, despite attempts by many scholars, as we will see in this book, no serious breakthrough has occurred to provide a strong model or a coherent theoretical exposition of this phenomenon. Until recently, explaining scientific change and progress dominated the agenda, particularly in the wake of the reception of Thomas Kuhn's popular theory of scientific revolutions as the best explanatory device for understanding the change and progress (despite Kuhn's protestations about the incommensurability of scientific paradigms) of the scientific enterprise.[3] Attempts to extend Kuhn's model (paradigm shifts brought about by scientific revolutions during the historical development of a scientific discipline) to model technological change only selectively explained it, because, unlike scientific change, technological change follows mostly a continuous and evolutionary trajectory. While technological change is essentially a continuous and cumulative process in societies that are characterized by continuity and relative stability, scientific change, on the other hand, tends to be a noncumulative and discontinuous process, as Kuhn and others have shown.[4]

The greatest stumbling block to understanding technology and its multifarious ways of development until recently was the prevalent misconception that technology was only "applied science," hence, technological change needed no separate explanation. Technology remained the handmaiden of science, for most science policymakers and historians of science (also, perhaps, for the intellectual community at large), until epistemic and cognitive attributes were located and identified for technology by mostly historians and sociologists of technology. The acceptance of technology as an autonomous (from science) intellectual enterprise has been the stepping-stone for the recognition of technology as a form of knowledge. Although technology is autonomous with respect to science and vice versa, neither is autonomous with respect to society and culture. Technology, thus, came to be recognized at the primary level as artifacts and at the secondary and tertiary levels as the social and cultural expression of a people or society, and the cognitive focus of the community of technologists and technology practitioners, broadly conceived. The new intellectual focus on technology is to look at it as a body of knowledge with its own internal dynamics of change and progress.

The claim that technology is merely applied science and that technological knowledge production is not possible without scientific knowledge is an ahistorical claim without any theoretical basis or empirical support.[5] Technology originated long before anything we can call science evolved.[6] The developmental dynamics of technology and science can be traced separately; however, the interdependence of each other has become more intense only during the twentieth century. Humans have been using tools and devices without understanding the theoretical basis of most of them. Humans made fire without knowing what caused combustion. The oxidation theory of combustion became available only in 1774 when oxygen was discovered.[7] Numerous ancient civilizations had used catapults, levers, bows and arrows, waterwheels, and numerous other technological artifacts without proper knowledge of the theoretical principles behind the workings of these artifacts.

Although archaeologists, historical anthropologists, and others have been sifting through the ruins of ancient civilizations such as the Egyptian, Mesopotamian, Chinese, Mayan, Aztec, and Greek to identify their technological wherewithal and other cultural artifacts with which they built their civilizations, no scholar has come up with a clear theoretical explanation of the intricacies of the technological advances these civilizations had experienced beyond certain generalizations.[8] Because of the daunting nature of the need to reconstruct the historical records and available artifacts, it may be almost impossible to create a cogent narrative history of the technological changes these civilizations underwent. Looking at the marvellous pyramids of Giza, we can readily admit that the Egyptians of antiquity had a breathtaking ability to quarry, move, and carve incredibly large stones. We can only speculate that their civil, mechanical, and hydraulic engineering knowledge was highly advanced. Yet, we are still at a loss to explain exactly how the Egyptians built these wondrous megaliths (or what Lewis Mumford calls "megamachines"[9]), because we do not have written records of their engineering practices or access to the tools, ships, and machines with which they built these structures.

Since technological change does not occur in a vacuum, but rather in a social context, it is extremely important to have a clear exposition of the social dynamics of the particular nation/society for the task at hand. Nonetheless, we do have fairly good historical accounts of the technological know-how, for example, of the ancient Chinese, Greeks, Romans, and Egyptians.[10] In an attempt to explain the dynamics of premodern technological change, Lynn White shows that Christianity played a central role in the technological change that had taken place during the Middle Ages in Europe.[11] White argues that unlike Eastern religions, Christianity celebrated the purported dualism between humans and nature, and legitimated the exploitation of nature as an act ordained by God.[12] White identifies "small things," such as the increased use of iron, the replacement of oxen by horses, the harnessing of various natural power, the introduction of the stirrup in warfare, and the modernization of the plow as the major causes for the rapid technological change in medieval Europe. However, White's views on technological change have been severely criticized by Hilton and Sawyer, who argue that White resorts to "technical determinism" and supports his case with a "chain of obscure and dubious deductions from scanty evidence about the progress of technology."[13] Therefore, it is important to avoid such pitfalls in interpreting ancient technological change without proper recourse to convincing empirical records. However, we may be on safer ground to look for the same conditions in modern times where better evidence and historical records are available for analyzing technological change, for developing appropriate models or theories, and for interpreting the process of change in the knowledge base of modern or modernizing societies and nations.

It is no surprise that bolder and more provocative historical accounts of technological change are available about the modern and early modern periods, perhaps because of our spatial and temporal proximity to the events and things. David Landes, for example, chronicles the unbinding of the Promethean gift of modern technology and the unleashing of wealth and prosperity in the West as a result of the Industrial Revolution.[14] Landes claims that the reasons for the spectacular rise in the prosperity of the West were attributable to its abiding faith in rational manipulation of the environment and to its particular institutional structures. Many other scholars have attempted similar sweeping interpretations of technological change. Abandoning his earlier claim that technological change is determined by the exigencies of capitalist forces and their agents of state power, David Noble now puts forward an omnibus theory that the mindless growth and expansion of the technological enterprise in America and the West is driven by the religious fervor of their technologists.[15] Noble postulates that technological change is driven by the religious "imaginings" of the creators of technology—the technologists— whether they live in the ancient or modern period. According to him "modern technology and religion have evolved together," a claim he makes by attributing agency exclusively to professed freemasons, millenarians, evangelicals, and other assorted religiously bent scientists and engineers.[16] There is no dearth for grand hypotheses and speculations about the causes and consequences of technological change. Yet, a coherent and empirically based theoretical account of this seemingly inscrutable phenomenon is still unavailable.

WHAT IS TECHNOLOGICAL CHANGE?

Although intuitive theoretical and empirical explorations on understanding technological change have been slow in coming, most intellectual disciplines have shown considerable interest in this phenomenon either directly or through implication. At the core of most contemporary social, political, economic, environmental, and ethical problems, processes and concerns have a deep technological component as a constituent or as a causal factor. Although most of the intellectual deliberations on these issues may not involve a serious discussion of the way technological change affects their concerns and issues, allusion is often made to technology's role in the genesis and future course of these concerns and issues. Obviously, this preoccupation is counterintuitive, because accounting for the growth of knowledge in itself, without a qualitative deconstruction of its various manifestations, would be problematic in accounting for the advances in human cultures and civilizations.[17] This residual accounting has to come up with an explanation for "progress" if the particular discipline or knowledge enterprise can only account for a piece of the larger "puzzle" of social change and progress, while whatever form of interpretation or explanation they are engaged in that corresponds to their particular ontological concerns.

Although the interest in technological change shown by most disciplines was not prompted by a desire to theorize and explain this phenomenon, but rather by a well-founded belief that the change in the technological knowledge base of societies corresponds to the increase in economic productivity and social change.[18] The belief that technological change can induce changes in people's material conditions of living, mostly for the good, is rather pervasive in almost all cultures. Social theorists, political economists, and Enlightenment and later modernist *philosphes* such as Marx, Engels, Saint-Simon, Smith, Ferguson, Hume, Condercet, and Turgot clearly comprehended the role technological advance played in bringing about social and economic change through the modernization project. Marx wrote about the way in which the rapidly evolving modern technology generated wealth and prosperity, albeit most of these benefit accruing to a minority in society. He showed that the means of production engendered by the new technological innovations dealt a fatal blow to the old relations of production during the precapitalist stage. Marx and Engels claimed that without "constantly revolutionizing the means of production," the bourgeoisie could not survive in the capitalist system of social production.[19] Although Marx and Engels lauded the "progressive" nature of capitalism (when contrasted to the regressive feudal system it replaced), they claimed that the relations and mode of production would eventually lead to the alienation of the workers (who according to Marx and Engels are the sole producers of value) from the means of production (technology). As Rosenberg correctly noted, Marx was a great student of technology who reflected on the dynamic nature of technological change (more about which will be discussed in chapters two and four) and its impact on social and economic change.

Although interest in technological change declined considerably after Marx, it picked up again during the early part of this century with the works of Veblen, Schumpeter, Usher, and Gilfillan (whose works, excluding Veblen, are analyzed in several of the subsequent chapters). Concurrently, the *Annales* School of intellectual historians, particularly Marc Bloch and Lucien Febvre, began an exciting genre of economic history by focussing on technological change. Although these analysts emphasized different aspects of technological change, particularly their emphases upon using certain idiosyncratic vocabularies for seemingly similar concepts, their ultimate interests were related to the process of change in the technological knowledge base of society. After a respite of a few decades, beginning in the 1950s, considerable interest in technological change was shown by two groups of economists. One group was interested in the ways by which modernization and economic development could be induced by technological change in the newly independent Third World nations through the transfer of modern (Western) technology to these "underdeveloped" states (for a detailed treatment of some of these issues, see chapter six). The other group consisted of neoclassical economists who were attempting to explain economic growth and wealth creation in Western nations after the onset of the Industrial Revolution. Robert Solow, Moses Abramovitz, Edwin Mansfield, Zvi Griliches, and other neoclassical economists were not primarily concerned about the process or the internal dynamics of technological change in itself, but rather in the mechanisms by which nations grow economically and individuals accumulate wealth as a result of the advances in productive enterprises spurred on by modern technology.[20] The productivity puzzle of not being able to account for the increase in economic growth after deducting the shares of productivity growth attributable to factors of production (capital and labor) led these neoclassical economists to the "discovery" of technological change as the missing link. That is, they serendipitously came to know that it was technological change that created much of the growth in income and wealth.[21] The predominant neoclassical approach of approximating the effects of technological change on productivity growth was undertaken by modelling the economy using a production function. The economic production function used for this analysis was a variant of an engineering production function of input/output relationship adapted for value data in monetary units. This process inevitably ended up in black boxing technological change in the productivity model (for more detailed analysis, see chapter four).

It appears that until the 1960s only economists paid serious attention to technological change, albeit in an indirect way. However, soon after the publication of Kuhn's *The Structure of Scientific Revolutions* scientific change received considerable attention, and unfortunately, technological change was treated by most scholars as an epiphenomenon of scientific change. With deftness and intellectual insight, some historians and sociologists of technology rescued technological change from the clutches of "applied science." Technology was treated as an autonomous (from science) enterprise with its own intellectual and epistemological issues pertaining to growth and change. Most of these analysts (as we will see in chapters two, three, four and five) proposed their own interpretations of technological change. They, in turn, developed models to represent and explain this phenomenon. These analysts worked within their own areas of inquiry or disciplinary matrices. A few among them attempted to transcend the boundaries of their disciplines by incorporating perspectives and theories from other disciplines. Theoretical constructs and metaphors such as "systems," "evolution," "social construction," and "actor-networks" showed considerable promise in explaining technological change than the production function model popularized by neoclassical economists. Despite being bold and pioneering, most of these models needed to evolve further in order to capture the complexity and dynamism of technological change, as will be argued in this book.

The objective of analyzing or modelling technological change tends to vary from discipline to discipline, although most analysts implicitly concede that technological change is a necessary condition for economic development and social change. Economists may be interested primarily in the role technological change plays in productivity growth and changes in economic structures and labor relations. Sociologists and anthropologists may be looking for the changes in the social structures, cultural practices, and social adaptation of various agents (workers, consumers, youths, and so on) as a discursive process that occurs within the ambit of reflexive modernization.[22] Because of the self-reflexive nature of modernity, the study of how technological change evolves due to the dialectical relationship between technology and society is an important topic for further investigation in its own right. Historians may tend to concentrate on chronicling the internal or external changes in the making of artifacts and their role in cultural and civilizational change. Philosophers may concentrate on the moral and ethical implications of technological change and progress, and develop methodological tools for technology assessment, or reflect on the normative dimensions of technology in human affairs. Many analysts from a variety of disciplinary backgrounds who see this phenomenon as a cumulative and gradual process use Darwinian evolutionary theory in varying degrees of intensity to model and explain the process of technological change. While some of these analysts use biological evolution in a mere metaphorical sense to analyze technological change, others treat technological change as isomorphous to biological evolution.

Despite caveats and warnings, most analysts across the disciplines accept technological change as a positive force in society and exhort the need to spur on technological innovations to solve many of society's problems. However, it is not necessarily the case that technological change fascinates only those interested in its effects on progressive social and economic change resulting in the improved well-being of humans. Analysis of technological change may be undertaken to understand the causes of technological failures and disasters. As some sociologists of technology strongly argue, we may learn more about technological change from studying failed technological innovations (see chapter three for further elaboration). It is not necessarily the case that technological change equally benefits all classes or groups in society. There are several thinkers who consider rapid technological change as a potentially dangerous development. According to them, modern technology has become an autonomous force that has acquired the momentum and direction with which it steers its own course of change and renewal. This "deterministic" force emanating from technological change is supposed to threaten human liberty and even survival. Thinkers such as Jacques Ellul and Langdon Winner argue that the uncritical acceptance of the rewards of modern technology in the form of progress and for accumulating goods and services to supposedly improve the standard of living would inevitably lead to the destruction of human freedom and liberty.[23] This happens, according to them, because human beings become captive to the "technological imperative." Although this facet of technological change is an important topic for further scrutiny, it is, however, beyond the scope of this book.

Interest in the phenomenon of technological change is an important topic for planners and policymakers in the developing world. This is the result of a near-consensus belief (based on empirical findings from the industrialized countries) that high rates of economic growth and productivity increase have some correlation to technological change.[24] An important causal linkage to the spectacular rise in the overall standard of living in the industrialized nations can be attributed to rapid technological change that these nations had experienced during and after the Industrial Revolution.[25] The ideology that technological change is a necessary and sufficient condition for economic growth and social reconfiguration of societies is deep-rooted in modern industrial cultures. The juggernaut of technological change as an open-ended process without a clear trajectory is an essential element of the modernization project embedded in the western ontology.[26] Therefore, investing in technological knowledge production is considered essential for the economic survival of all modern and modernizing societies. It has become an inductive thinking among the protagonists of modernization that without continuous economic growth spurred on by technological innovations the modern sociotechnological system might collapse.[27]

For the newly decolonized developing nations, according to the pundits of international development, the easiest path to join the modernization bandwagon of the developed nations was to acquire modern technology to modernize their "underdeveloped" economies. The leaders of these newly independent nations followed essentially two variants of the modernization project: the free market capitalist and the planned statist growth models with modern technology acting as the common denominator. Since these developing nations lacked the technological and scientific wherewithal to mobilize the forces required for economic development from within, transferring technology from the industrialized world under an array of developmental assistance programs became the norm.[28] Although technology transfer is considered as an important step in inducing technological change in the developing world, no explicit theory or model is available to implement this program. A major reason for the failure of most technology transfer programs was the lack of understanding on the part of the planners and policymakers as to what constitutes technological change and how it can be achieved or induced in a nation/society. Such knowledge is essential for bringing about the change. One of the objectives of this book in explaining the process of technological change is to understand the mechanism of technology transfer (see specifically, chapters six, seven, eight and nine). The so-called Green Revolution in Indian agriculture will be used as the empirical focus to substantiate this claim.

Given that interest in technological change varies according to the disciplinary and individual objectives of the analysts, it should come as no surprise that there is no consensus on a consistent theoretical explanation of this phenomenon. Although it can be explained in a causal or functional or intentional manner, explanations provided by each of these by themselves may not give a complete picture of this complex phenomenon. Perhaps a synthetic or syncretic approach that combines these three steps might be more appropriate. Likewise, it may be an epistemologically risky exercise to look for a simple and consistent definition of technological change. This risk is primarily due to the lack of a consistent and clear definition of technology itself.[29] Without a clear understanding of what this phenomenon is all about, it would be rather presumptuous to offer a clear definition! Because of this vicious circularity, most works on the subject circumvent the need for providing a clear definition.[30] Most analysts, unable to articulate it, forego this part and go on to explain the impact of technological change on humans and society, or explain what its structural components are, or offer models based on case studies.[31] According to Jon Elster, who dares to define it, technological change is a "rational goal-directed activity" which is seen "as the best innovation among a set of feasible changes."[32] This minimalist definition basically affirms that the changes that accrue to a production system (any activity that is conducted to solve a problem) occur as a result of the changes in the knowledge of production concerned with solving technical tasks.

With the above concerns, criticisms, caveats, and available definitions in mind, one may "define" technological change (in an intentional sense) as a temporal and cumulative process that increases the ability or potential of a people or society to solve their social, economic, environmental, and other everyday existential problems or needs. Technological change then manifests (in a causal sense) as a problem-solving ability (of the agents), which often becomes conspicuous in tangible and material (artifactual) forms. However, all abilities and potentials come closer to a cognitive process, which is reflected in the knowledge attributes of the agents. Therefore, technological change is, essentially, a process of knowledge change — a cognitive process involving the agents (this concept will be elaborated in chapters seven, eight, and nine).

A more rigorous definition of technological change is that it is the outcome of the activities that humans engage in, through their individual efforts or collective organizational structures (such as firms, households, or any other institutional agents or social units of production and reproduction), to optimize their resources (existing knowledge, energy, time and materials, and so on) subject to constraints imposed by their own limitations. These qualifying factors can come from within the individual or the collective entity that she belongs to (or part of) and also from the environment. Once the limits have been ascertained, technological change, therefore, may be analyzed as a problem-solving activity (an elaboration of this concept is attempted in chapter six, seven and nine). However, one must keep in mind that technological change is not the end result of a completely optimizing action or behavior of the agents. The problems generated by "technological imbalances," "technological bottlenecks," "technological disequilibria," and other such puzzles act as "focusing devices" to find the appropriate solutions. The puzzles are solved by any number of methods, such as through meticulous research and development efforts, innovations on shop floors and in the fields and farms, encouraging new inventions, and other such problem-solving methods.[33] Although the problems appear to be functional in nature, the solutions must be achieved by looking at the problems from an intentional perspective. That is, the functional issues need to be translated into deliberate designs, production methods, and other activities. The translation involves internalizing the problem through learning processes, such as learning by doing, learning by using, learning by scaling, and so on. That is why technological change is looked at in the functional sense as the outcome of a problem-solving activity conducted in an intentional manner.

Subsequently, technological change should be viewed as the unfolding of the invention, innovation, development, transfer, and diffusion of a technological system, which may contain a single or a multiplicity of these interacting components or vectors. Technological change, simply put, is the historical account of a technology's birth, growth, and sometimes demise. Although technological change can be characterized as the combined effect of the five stages mentioned earlier,[34] it is not necessarily the case that these components have to follow the same sequence in every case of technological change, nor is it mandatory to have all of these components in every instance of technological change. These components are conceptually distinct; however, it is empirically often difficult to tell when one ends and the other begins. The best approach to understanding technological change is to treat the recently mentioned components that make up this phenomenon, not as a chronologically occurring sequence, but instead, as forces (vectors) operating in an interactive-dialectical manner, mediated by social forces and interests, following an evolutionary trajectory. What technological change boils down to is that it is a process of knowledge change of the actors' (or their collective endeavors like social systems) manifested in the form of technological knowledge related to problems of production, distribution, communication, and other needs. Technological change is knowledge change, a theme taken up in greater detail in chapter nine.

TECHNOLOGY AS KNOWLEDGE AND TECHNOLOGICAL CHANGE

Establishing the intellectual autonomy of technology from science (that is, technology is *not* "applied science") led to spectacular changes in the way technology was perceived by the science and technology studies community. A major outcome of this development was the interest shown in understanding what is technological change and how it is effected and induced, and also if there is anything that can be identified as technological progress. Like scientific knowledge, the knowledge associated with the practice of technology, broadly conceived, gained significance because how new technological inventions and innovations emerge cannot be understood by looking at the artifactual realm of technology. How technology transfer works cannot be learned by looking at the spatial transference of seemingly tangible material artifacts such as machines, instruments, and equipment from one geographical locality to another. One needs to go further than the material realm to understand how a dynamic component of technological change, for example, technology transfer or technological innovation, takes place. One needs to reveal the epistemic basis of technology to unravel the mysteries of these technological activities.

Attributing a knowledge component to technology is very important to understand technological change. The epistemic significance of technology stems from its clear dependence and interfacing with society and culture. Analysts of technology, particularly historians of technology, have shown that "technology as knowledge" is a useful concept for understanding technological change.[35] A major objective of this book is to expand further on this concept and to show that technological change can be understood best as knowledge change. I will show this in chapters seven, eight and nine using the Green Revolution in Indian agriculture as the empirical focus, within the overall frame of technological change as problem solving.

What we will look for in the Green Revolution case study is not just the history of an agricultural revolution by analyzing the typology of various artifacts that had evolved during the process of this technological change, but rather the process of knowledge change that was experienced by the agents. The Green Revolution, as we will see in chapters six and seven, is a controversial event because of the adverse environmental and social problems that erupted while the authorities were trying to solve the problem of low agricultural productivity in India through an apparent technological fix. The reconstruction of the technological change in the Green Revolution will not be undertaken as a classic case study in the history of technology frame, for the knowledge and practices associated with agriculture are as much in the scientific realm as they are in the technological realm. It straddles both science and technology. Following Michel Foucault's footsteps, what we will do is to "focus on the cognitive significance of 'natural-historical' understandings, whether of divine or human creation"[36] while constructing the network of elements and actors in the Green Revolution. Following John Pickstone[37] (who uses Foucault's historiographic method to analyze history of technology), by expanding on the notion of technology as culture through "the study of knowledge-practices and practical-knowledges" and integrating "the cognitive and the social aspects" of the technological change in the Green Revolution, we will reveal the epistemic basis of technology.

A correct understanding of technological change has great policy implications as we argued earlier in the context of the developmental agenda of the modernization project. The idea of developing a coherent and comprehensive model of technological change is not only to understand it, but also to intervene in society in order to bring about "desirable" social change as much as possible, despite the contingencies involved in the actual process of change itself. But as Wiebe Bijker forcefully argues, the contingent nature of technological development should not unravel the project of constructing stable theoretical "frames" to explain technological change.[38] Bruno Latour argued that the best way to learn about science and technology is to look at the process of technoscientific knowledge production in action; that is, to follow the actors who create the knowledge while engaged in the social construction of the technological "actor network."[39] However, one needs to know which actors to follow for this suggestion to become a meaningful enterprise. The most important analytical tool that Bruno Latour, Michel Callon, John Law, Wiebe

Bijker, Trevor Pinch, and Thomas Misa, among other analysts of technological change, propose to understand this process is to first construct the sociotechnological network through which the knowledge production unfolds.[40] But as Misa correctly points out, when one delves deeper to unravel the intricacies of technological change, one needs to show how change is effected through mesolevel agents (institutional actors and so on) whereby the process of "social shaping of technology and the technological shaping of society" becomes apparent.[41]

Thus, it is hoped that by opening the black box of technology we can identify which actor(s) to follow, so that the ensuing technological change can be successfully represented through a practical-reflexive mode of theorizing as explained in more detail in chapter seven. The objective of this book is to follow the lead provided by Bruno Latour, Trevor Pinch, Wiebe Bijker, Thomas Misa, and other analysts of technological change mentioned earlier to represent and intervene in this phenomenon. However, as we will see in this book following the actors is only a beginning and more surprises will be in waiting once we open the black box.

ORGANIZATION OF THE BOOK

The objective of this book is to present a comprehensive theoretical and empirical account of contemporary technological change by charting a new trajectory in technology studies. This is undertaken with the aim of initiating a new "research tradition" in STS to show that like scientific change, in studying technological change one should also have an ontology, methodology, and epistemology to guide the construction of the analytical framework. The ontology should consist of some fundamental beliefs and expectations of the technology studies community; that normative issues of what should be the "object of inquiry" of technological change must not be disregarded in theory and model building.[42] The methodology should be founded on solid foundations of proper empirical studies. Finally, the epistemological assumption is founded on the assumption that "technology is knowledge" and the idea of knowledge change of the actors is the key to understanding technological change. Although these three epistemic goals will not be spelt out in every instance in my investigation, their influence can be detected in the overall spirit of the book. What is being implied here is that the objective of studying technological change is to intervene as much as to represent in an open-ended investigation of the epistemic significance of technology as knowledge.

As we saw in the previous two sections, a detailed theoretical account of this phenomenon has been presented in this chapter as a backdrop for the whole book. I have provided a clear definition of technological change because of the fact that despite innumerable works available in the literature on various aspects of technological innovation, transfer and diffusion, and other extant aspects of the process of technological change, it is still a poorly understood concept. Therefore, it is hoped that these explanations and the definition of technological change will provide the central focus for laying the intellectual foundation of the book. In chapters two through five, I will provide a comprehensive, critical analyses of how technological change is conceptualized and problematized in the various participating disciplines and areas of interest of STS such as history, sociology, economics, and neo-Schumpeterian evolutionary studies of technology. The objective of this analysis is not to present an encyclopaedic description of how technological change is presented in these disciplines, but rather to show, in a selective manner, how it is being conceptualized theoretically and presented empirically. A major criterion used for the selection of various models and approaches of technological change from these disciplines and areas of inquiries for my critical analyses is their heuristic potential for modelling/explaining other instances of technological change (i.e., other than the empirical cases used by these analysts).

In chapter six, I will show that technological change can be understood as a problem-solving activity using the so-called Green Revolution in Indian agriculture as the empirical focus. In chapter seven, I will argue that "practical reflexivity" can be an effective heuristic for theorizing technological change using the case study of the Green Revolution for empirical support. In chapter eight, I will use selected prominent models from history, sociology, economics, and neo-Schumpeterian evolutionary studies presented in chapters two, three, four, and five to account for the technological change in the Green Revolution. The objective of this exercise is to see if these models can provide a consistent explanation of this phenomenon and, in addition, to verify whether or not they converge. Again, in a Popperian spirit, the objective is not to prove that these models/theories are true or false (which is impossible anyway), but rather to show their problem-solving effectiveness.

In chapter nine, I will argue that technological change is best understood as a cognitive process, that is, technological change is essentially knowledge change. In order to do this, attempts will be made to bring together some of the relevant ideas from the models and theories discussed earlier. A checklist of what factors and features (Eleven Theses on Technological Change) a comprehensive model of technological change ought to have is presented in this chapter using the Green Revolution as empirical support to strengthen the arguments. An important objective of this book is to look at the technology policy implications of analyzing technological change as a problem-solving activity, which should be understood as a contingently shaped social-historical process involving the knowledge change of the actors. This is undertaken to understand the process of technology transfer, an objective that is intended to help those interested in developing viable models of technology transfer and so on among other practical reasons for studying technological change. Needless to say, the objective of studying technological change should not be restricted to practical considerations. Rather, given the importance accorded to technological change by different social agents, regardless of their intentions, any intellectual enrichment of our understanding of it should be deemed a worthwhile endeavor. Also, it is hoped that by providing an effective intellectual groundwork, this book would stimulate further theoretical and empirical explorations on technological change.

ENDNOTES

1. The utopian views on modern technology are held mostly by the "cyberlibertarians" who claim that the overthrow of matter by the quanta of bits, bytes, photons, and so on would unleash the real laissez faire capitalism on our planet bringing with it unlimited individual freedom and wealth. Cyberlibertarians such as George Gilder, *Microcosm: The Quantum Revolution in Economics and Technology* (New York: Simon & Schuster, 1989), and Esther Dyson, *Release 2.0: A Design for Living in the Digital Age* (New York: Broadway Books, 1997), claim that the "Third Wave" revolution of advanced digital communication systems of the late twentieth century would bring with it unlimited wealth and individual freedom through an unimpeded informational capitalism of the emerging computerized civilization. Dystopian critiques of technology such as Langdon Winner, "Technology Today: Utopia or Dystopia," *Social Research* 64 (1997): 989–1017, *The Whale and the Reactor: A Search for Limits in an Age of High Technology* (Chicago: University of Chicago Press, 1986), and *Autonomous Technology: Technics-out-of-Control as a Theme in Political Thought* (Cambridge, MA: MIT Press, 1977), on the other hand, warns that these new technologies would destroy human freedom, jobs, and the very existence of humans as members of social units such as families and communities. Although not a focus of this book, more thoughts on how to transcend technological pessimism or unabashed technological optimism can be found in Govindan Parayil, "Transcending Technological Pessimism: Reflections on an Alternative Technological Order," *Working Papers in the Social Sciences, No. 43,* Division of Social Science, Hong Kong University of Science and Technology (28 December 1998).

2. According to Robert Fox, "Introduction: Methods and Themes in the History of Technology," in *Technological Change: Methods and Themes in the History of Technology,* Robert Fox, ed. (Amsterdam: Harwood Academic Publishers, 1996), 1–15, this conference was only a sequel to an earlier path-breaking conference on "The structure of scientific change" that was held at Oxford in 1961. Fox's edited volume contains some of the important and topical papers that were read to the 1993 conference. A notable contributor to the 1961 conference was Thomas Kuhn, who presented glimpses of his now famous theory of scientific change in his paper "The function of dogma in scientific research." The 1993 Oxford conference on technological change was a clear indication of the importance that technological change has been receiving in the 1990s, and it is in many ways as significant as the conference on scientific change that took place in 1961. It is, however, important to mention that despite the presence of most of the major scholars on technological change, whose scholarship spans many disciplinary fields, the theme of the "Technological Change" conference was limited to looking at the methods and themes from the perspectives of the history of technology.

3. Thomas S. Kuhn, *The Structure of Scientific Revolutions,* 2nd ed. (Chicago: University of Chicago Press, 1970). It is not meant that other competing theories of scientific change are not available to explain scientific change. In fact, Kuhn's theory has been seriously challenged by many scholars as being "relativistic," "irrational," and so on. Nevertheless, it is certainly true that Kuhn animated the discussion of scientific change from being the purview of a small group of philosophers of science and helped to carry the discussion to a larger intellectual group we call the STS community. Besides Kuhn, scholars such as Karl Popper, Imre Lakatos, Larry Laudan, Dudley Shapere, and a few others have offered alternative models or explanations of scientific change in the philosophy of science tradition. Being not the focus of my book, however, I will not go into these details here.

4. Although fundamentally new ways of doing things to solve a problem in production, communication, or other domains of technological application may appear to be revolutionary in nature, doing things and solving problems in new ways do not constitute technological revolution as we will show in this book. What has been achieved is better, easier, faster, or cheaper ways of doing or solving a problem that has been identified and solved, in most cases, using a previous vintage of technologies. What technological change essentially means is a new problem-solving culture (for an extensive treatment of this topic, see chapter six). This is contrary to scientific change, which is the discovery of new explanations and interpretations of the workings of nature and natural phenomena which previous theories failed to show adequately. Therefore, scientific change involves, invariably, the overthrow of old theories in favor of theories and models of nature and natural processes of better explanatory power and problem-solving effectiveness, while for technological change the matter is a different one, as we will find out from this book.

5. See Walter G. Vincenti, "Technological Knowledge without Science: The Invention of Flush Riveting in American Airplanes, ca. 1930–ca. 1950," *Technology and Culture* 25 (1984): 540–576; Derek J. de Solla Price, "Is Technology Historically Independent of Science? A Study in Statistical Historiography," *Technology and Culture* 6, no. 4 (1965): 553–568; George Wise, "Science and Technology," *OSIRIS,* 2nd Series, 1 (1985): 229–246; and Rachel Laudan, "Cognitive Change in Technology and Science," in *The Nature of Technological Knowledge: Are Models of Scientific Change Relevant?* Rachel Laudan, ed. (Dordrecht: D. Reidel, 1984), 83–104.

6. For a persuasive argument, see George Basalla, *The Evolution of Technology* (Cambridge: Cambridge University Press).

7. See Kuhn, *Structure.*

8. According to Alois Huning, "Homo Mensura: Human Beings Are Their Technology—Technology is Human," in *Research in Philosophy and Technology,* Volume 8, Paul T. Durbin, ed. (Greenwich, CT: Jai Press, 1985), 9–16, 11, the pioneering work on the "anthropology of technology" was first done by Ernst Kapp in his *Grundlnien einer Philosophie der Technik* ("Foundations of a Philosophy of Technology," 1877), in which Kapp uses the term *Organprojektion* (organ projection) to refer to technology. Robert F. G. Spier, *From the Hand of Man: Primitive and Preindustrial Technologies* (Boston: Houghton Mifflin, 1970), also attempts an anthropological explanation of technological change of primitive and ancient "man," but fails to lead us to anything new other than obvious generalizations. However, Bruce Mazlish, *The Fourth Discontinuity: The Co-evolution of Humans and Machines* (New Haven, CT: Yale University Press, 1993), provides a more plausible explanation for the development of technology, from primitive tools to advanced modern computers. Mazlish presents a revolutionary hypothesis that technology and human cultures evolve in a coevolutionary manner. Simply put, Mazlish, *Fourth Discontinuity,* 233, claims that humans are a continuum with machines they have created, that is, "tools and machines are inseparable from evolving human nature." Although interesting, I will not pursue this line of argument in explaining technological change as knowledge change, lest one may get trapped on a one-way track of sociobiological explanation of technological change. However, to be fair to Mazlish, I must admit that he does not knowingly endorse sociobiology as a basis for explaining the development of technology.

9. Lewis Mumford, *Technics and Civilization* (New York: Harcourt Brace Jovanovich, 1963).

10. For the history of Chinese technology, see Joseph Needham, *Science and Civilization in China,* Volumes 1–7 (Cambridge: Cambridge University Press, 1954–1988). For the Greek, Roman and Egyptian technology, see Geoffrey E. R. Lloyd, *Greek Science After Aristotle* (New York: Norton, 1973); John W. Humphrey, John P. Oleson, and Andrew N. Sherwood, eds., *Greek and Roman Technology: Annotated Translation of Greek and Latin Texts and Documents* (London: Routledge, 1998); and L. Sprague de Camp, *The Ancient Engineers* (New York: Ballantine Books, 1974).

11. Lynn White, Jr., *Medieval Technology and Social Change* (Oxford: Oxford University Press, 1962).

12. In fact, White lays the blame for the acute environmental crisis that the western world has been facing squarely on Christianity, for its legitimizing the dualism between "man" and nature. This dualism resulted in the belief that humans are outside of nature and that there is no contradiction in exploiting nature at any cost for the purported betterment of "man" in the name of God. See Lynn White, Jr., "The Historical Roots of Our Ecological Crisis," *Science* 155 (10 March 1967): 1203–1207.

13. Rodney H. Hilton and P. H. Sawyer, "Technical Determinism: The Stirrup and the Plough," *Past and Present* 24 (1963): 90–100, 90, found White's claim of the role that the stirrup had played in the formation of feudalism as mere assertion that lacks historical credibility. Richard Holt, "Medieval Technology and the Historians: The Evidence for the Mill," in *Technological Change,* Robert Fox, ed., 103–121, is also critical of White's assertion that the Middle Ages was a period conducive to great technological innovations. Holt argues that because of the lack of a reward system for innovators in a period dominated by craft practices and small-scale production conducted for local markets, the idea of a proliferation of technological inventions does not sound very convincing. Bert Hall, "Lynn White's *Medieval Technology and Social Change* After Thirty Years," in *Technological Change,* Robert Fox, ed., 85–102, on the other hand, attempts a passionate defense of *Medieval Technology and Social Change* and rejects Hilton and Sawyer's criticism as "vitriolic." Hall, "Lynn White's *Medieval,*" 87, argues that White is one of the greatest historians of technology ever, who "saw himself as the intellectual offspring of the *Annales* school founded by Marc Bloch and Lucien Febvre."

14. David Landes, *The Unbound Prometheus: Technological Change and Industrial Development in Western Europe* (Cambridge: Cambridge University Press, 1969).

15. David F. Noble, *The Religion of Technology: The Divinity of Man and the Spirit of Invention* (New York: Knopf, 1997).

16. Noble, *Religion of Technology,* 5.

17. The obvious assumption being that technological advance is the key factor in the material well being of humans and the advancement of civilizations.

18. One important manifestation of these changes was the dramatic improvement in the standard of living of the people (mostly in the western industrialized nations) who experienced rapid technological change during the past 100–150 years.

19. Karl Marx and Friedrich Engels, *Manifesto of the Communist Party* (Chicago: Encyclopaedia Britannica, 1952).

20. For some representative works of some of these neoclassical economists, see Robert Solow, "Technical Change and the Aggregate Production Function," *Review of Economics and Statistics* 39 (1957): 312–320; Moses Abramovitz, "Resource and Output Trends in the United States since 1870," in *The Economics of Technological Change: Selected Readings,* Nathan Rosenberg, ed. (Harmondsworth, UK: Penguin Books, 1971),

320–343; Zvi Griliches, "Hybrid Corn and the Economics of Innovation," in *The Economics of Technological Change,* Rosenberg, ed., 212–228; and Edwin Mansfield, *Technological Change* (New York: Norton, 1968).

21. For more on this, see Govindan Parayil, "Economics and Technological Change: An Evolutionary Epistemological Inquiry," *Knowledge and Policy* 7, no. 1 (1994): 79–91.

22. For further explication of the notion of reflexive modernization, see Anthony Giddens, *Modernity and Self Identity* (Stanford, CA: Stanford University Press, 1991), and *The Consequences of Modernity* (Stanford, CA: Stanford University Press, 1990); Scott Lash, "Reflexive Modernization: The Aesthetic Dimension," *Theory, Culture and Society* 10, no. 1 (1993): 1–24; and Ulrich Beck, *Risk Society: Towards a New Modernity* (London: Sage Publications, 1992). Beck sees reflexive modernity as coextensive with his notion of a "risk society." According to Beck, *Risk Society,* 19, what characterizes the latter is the intensive utility of technological knowledge in molding the industrial society that is being pervaded by a proliferation of negative externalities of "technoscientfically produced risks." While technological change plays a key role in the production of wealth in industrial societies, the same phenomenon is implicated in the production and distribution of acute ecological and industrial problems.

23. Further elaboration of these and related ideas of these philosophers of technology can be found in Jacques Ellul, *The Technological Society* (New York: Knopf, 1964), and Langdon Winner, *Autonomous Technology; Whale and the Reactor;* and "Technology Today."

24. For a classic treatment of this case, see Lester B. Lave, *Technological Change: Its Conception and Measurement* (Englewood Cliffs, NJ: Prentice-Hall, 1966).

25. As will be shown in chapter four, despite their shortcomings in explaining how technological change actually takes place, Robert Solow's "Technical Change," and other neoclassical economists' bold conjecturing that economic growth is predominantly a byproduct of technological change that nations and states undergo, is noteworthy. In fact, their empirical studies show a clear correlation between technological change and economic growth.

26. For an excellent treatment of this ontological position of the West, see George Kateb, "Technology and Philosophy," *Social Research* 64 (1997): 1224–1246.

27. The system dynamic modeling of the global economy conducted by Donella Meadows, Dennis Meadows, Jorgen Randers, and William Behrens, *Limits to Growth: Confronting Global Collapse* (New York: Universe Books, 1972), forewarned about the possibility of this collapse. Their cybernetic modeling shows the need for a continuous positive feedback of technological innovations and natural resources for the survival of the modern economic system. Their neo-Malthusian policy prescription was to radically reduce population growth, consumption, and pollution rates.

28. It is beyond the scope of this book to delve deeply either into the theoretical foundations of modern developmentalism or a postmortem analyses of the shortcomings of the modernization project that failed to lift much of the humanity from poverty and deprivation. Although there are numerous works on the subject, I want to point out one illuminating work by Arturo Escobar, *Encountering Development: The Making and Unmaking of the Third World* (Princeton, NJ: Princeton University Press, 1995), as a starting point for further reading.

29. That does not mean that there is paucity for definitions. Philosophers of technology normally begin their discourse on technology by first offering definitions. Ernst Kapp defines technology as "organ projection" (see Alois Huning, "Homo Mensura"). Martin

Heidegger, *The Question Concerning Technology and Other Essays* (New York: Harper & Row, 1977), defines the essence of technology as "revealing." While revealing the truth about nature by actively intervening in natural forces, Heidegger argues, what humans actually do, in a self-reflexive way, is to reveal the truth about themselves. Jacques Ellul, *Technological Society,* defines technology as the totality of human rationality to achieve absolute efficiency at doing things. For modern philosophers of technology such as Carl Mitcham, *Thinking Through Technology: The Path Between Engineering and Philosophy* (Chicago: University of Chicago Press, 1994), technology is what humans engage in, in an "unthinking" way, to make and use artifacts. For Joseph Pitt, " 'Style' and Technology," *Technology In Society* 10 (1988): 447–456, technology is "humanity at work." Although these philosophical reflections on technology do tell us something about technology, what they actually reveal is the ontological predisposition of the philosophers themselves. Unfortunately, they do not help us understand how technological development actually takes place and how change occurs in the technological knowledge base of nations and societies.

30. For example, in the works of such analysts of technological change such as Chris de Bresson, *Understanding Technological Change* (Montreal and New York: Black Rose Books, 1987), and Charles Susskind, *Understanding Technology* (Baltimore: Johns Hopkins, 1973), one cannot find a clear definition of technological change, nor can one get a clear explanation of what this phenomenon is all about, which is left to the readers to make out. In a similar vein, Maureen McElvey, *Evolutionary Innovations: The Business of Biotechnology* (Oxford: Oxford University Press, 1996), studies the development of molecular biology–based biotechnology pharmaceutical industry by looking at technological change as just innovation in the Schumpeterian frame.

31. This is not to claim that definitions are never found in the literature. On the economics of technological change in the neoclassical genre, one can find such minimalist or meaningless definition of technological change as capital-labor ratio in a production function framework. Thus technological change may be defined as biased in favor of either capital or labor, or neutral if the ratio is one.

32. Jon Elster, *Explaining Technical Change: A Case Study in the Philosophy of Science* (Cambridge: Cambridge University Press, and Oslo: Universitetsforlaget, 1983), 9.

33. For more on this conceptualization of technological change, see Nathan Rosenberg, "The Direction of Technological Change: Mechanisms and Focusing Devices," *Economic Development and Cultural Change* 18, no. 1 (1969): 1–24, and *Perspectives on Technology* (New York and Cambridge: Cambridge University Press, 1976); and John F. Hanieski, "The Airplane as an Economic Variable: Aspects of Technological Change in Aeronautics, 1903–1955," *Technology and Culture* 14, no. 4 (1973): 535–552.

34. For further treatment of the structural components of technological change, see Lynwood Bryant, "The Development of the Diesel Engine," *Technology and Culture* 17, no. 3 (1976): 432–446, and Thomas P. Hughes, "The Development Phase of Technological Change," *Technology and Culture* 17, no. 3 (1976): 423–431.

35. Some of these notable analysts who pioneered the concept of "technology as knowledge" are: Edwin Layton, "Through the Looking Glass; or News from Lake Mirror Image," in *In Context: History and the History of Technology,* Stephen H. Cutcliffe and Robert C. Post, eds. (Bethlehem, PA: Lehigh University Press, 1989), 29–41, "Technology as Knowledge," *Technology and Culture* 15, no. 1 (1974): 31–41, and "Mirror-Image Twins: The Communities of Science and Technology in 19th-Century America," *Technology and Culture* 12 (1971): 562–580; Walter G. Vincenti, *What Engineers Know and How*

They Know It (Baltimore: Johns Hopkins, 1990), and "Technological Knowledge"; Nathan Rosenberg and Walter G. Vincenti, *The Britannia Bridge: The Generation and Diffusion of Technological Knowledge* (Cambridge, MA: MIT Press); Price, "Technology Historically"; Eugene S. Ferguson, "The Mind's Eye: Nonverbal Thought in Technology," *Science* 197 (26 August 1977): 827–836; and Rachel Laudan, "Cognitive Change in Technology."

36. John V. Pickstone, "Bodies, Fields, and Factories: Technologies and Understanding in the Age of Revolutions," in *Technological Change,* Robert Fox, ed., 51–61, 53.

37. Pickstone, "Bodies, Fields," 61.

38. Wiebe E. Bijker, *Of Bicycles, Bakelites, and Bulbs: Toward a Theory of Sociotechnical Change* (Cambridge, MA: MIT Press, 1995).

39. Bruno Latour, *Science in Action: How to Follow Scientists and Engineers Through Society* (Cambridge, MA: Harvard University Press, 1987).

40. Latour, *Science in Action;* Michel Callon, "Society in the Making: The Study of Technology as a Tool for Sociological Analysis," in *The Social Construction of Technological Systems: New Directions in the Sociology and History of Technology,* Wiebe E. Bijker, Thomas P. Hughes, and Trevor Pinch eds. (Cambridge, MA: MIT Press, 1987), 83–106; John Law, "Technology and Heterogeneous Engineering: The Case of Portuguese Expansion," in *Social Construction,* Bijker, et. al., eds., 111–134; Trevor Pinch and Wiebe E. Bijker, "The Social Construction of Facts and Artifacts: Or How the Sociology of Science and the Sociology of Technology Might Benefit Each Other," in *Social Construction,* Bijker, et al., eds., 17–50; Wiebe E. Bijker, "The Social Construction of Bakelite: Toward a Theory of Invention," in *Social Construction,* Bijker, et al., eds., 159–187, and *Bicycles, Bakelites;* and Thomas J. Misa, "Retrieving Sociotechnical Change from Technological Determinism," in *Does Technology Drive History: The Dilemma of Technological Determinism,* Merritt Roe Smith and Leo Marx, eds. (Cambridge, MA: MIT Press, 1994), 115–141.

41. Misa, "Sociotechnical Change," 141.

42. See Larry Laudan, *Progress and Its Problems: Towards a Theory of Scientific Change* (Berkeley: University of California Press, 1977). It is duly acknowledged that Laudan's problem-solving approach to scientific change did influence my thinking on technological change, despite the lack of any methodological congruence between conceptualizing technological change and scientific change, as argued above and in Chapters 2 and 9.

2

Historical Models

A HISTORIOGRAPHIC NOTE

The idea that technological change is a historically contingent social and cultural process is well-known. Technological change being a temporal and largely a cumulative process, historical models can provide rich descriptions of the process of change in the technological knowledge base of nations and societies. Despite these promises, professional historians were late in according technology and the dynamic process of technological change their proper role in explaining the rise of the Enlightenment-driven project of modernization and the ensuing economic-social reconfiguration of societies and nations that led to the evolution of the modern industrial civilization. Edward Constant captures this sentiment appropriately by stating that "[T]his lacuna in historical understanding is symptomatic of history's broader failure to confront the process of technological change."[1]

While modern historians were reluctant to give technology its proper due in historical narratives, this was not based on any historical precedents or historiographic reasoning. It was simply due to professional historians' disdain to acknowledge that there is anything called a history of technology that is of intellectual significance. This was further compounded by the attitude of most historians of science who did not recognize the intellectual autonomy of the discipline of history of technology, as separate from history of science. They saw technology as simply "applied science." It was only recently that mainstream historians began to recognize history of technology as a separate disciplinary field. The intellectual focus of history of technology, in general, looks at technology as a form of knowledge, and is concerned about growth of knowledge as the central issue of technological change, more details about which will follow later in this chapter.

On a related note, the historical frameworks or mode of emplotments of "metahistorians" and historical sociologists can be employed as useful heuristics for developing rudimentary models and narrative accounts of technological change. The conceptual schemes and theoretical frameworks of Max Weber, Karl Marx, Oswald Spengler, and Arnold Toynbee, among others who reflected on social, cultural, and civilizational change, can be adapted to provide useful narrative accounts of technological change (more about these towards the end of this chapter). However, only a few scholars have attempted to provide detailed accounts of technological developments using these metatheorists' ideas and narrative modes.[2] A reformulation and application of the Marxian method of dialectical change to account for technological change is noteworthy in its own terms. Also, as Nathan Rosenberg observes, "A major reason for the usefulness of Marx's framework for the analysis of social change was that Marx was himself a careful student of technology."[3] However, professional historians, particularly historians of technology, have yet to pay serious attention to Marx's deliberations on the history of technology and his reflections on the nature and future course of technological change, more than a century after his works appeared.[4]

A key aspect of Marx's historical approach to technological change was his claim that technological inventions and innovations occur not as a result of the efforts of isolated heroic individuals, but as the outcome of collective social processes. In Volume I of *Capital,* Marx provides detailed descriptions of the development, working, and change of complex industrial machinery in large-scale manufacturing, while complaining that a relevant social history of industrial technologies has yet to be written. Marx's observations regarding the importance of the role that the history of technology plays in understanding social change are amply evident from the following passage from *Capital:*

> A critical history of technology would show how little any of the inventions of the eighteenth century are the work of a single individual. As yet such a book does not exist. Darwin has directed attention to the history of natural technology, i.e. the formation of the organs of plants and animals, which serve as the instruments of production for sustaining their life. Does not the history of the productive organs of man in society, of organs that are the material basis of every particular organization of society, deserve equal attention? And would not such a history be easier to compile, since, as Vico says, human history differs from natural history in that we have made the former, but not the latter? Technology reveals the active relation of man to nature, the direct process of the production of his life, and thereby it also lays bare the process of the production of the social relations of his life.[5]

Rosenberg rightly points out that these observations of Marx are a "prolegomenon to a history of technology that still remains to be written."[6] Marx viewed technological change in evolutionary terms, an indication of Darwin's influence, although he saw social change as discontinuous. Marx's historiographic approach was one that used historical materialism which "emphasized the interactions and conflicts of social classes and institutions, not individuals."[7] However, subsequent historians of technology, to a large extent, abandoned the larger social substrate of the "inventional" and "innovational" processes of technological change, and, instead, went back to such approaches as the heroic individual theory of technological change.

Recently, however, historians of technology have formulated a distinct historiographic method that is specific to technology that provided intellectually stimulating concepts and models of technological change, as opposed to general historians and historians of science who treated technology as simply "applied science."[8] Although scientific change has become a thriving research program, particularly since the enthusiastic reception of Thomas Kuhn's model in which scientific change and progress occur when the reigning scientific paradigm is overthrown in an intellectual revolution, technology is still waiting for its Thomas Kuhn.[9] Nevertheless, as a prelude, one can feel comfortable that the belated recognition of the intellectual autonomy of technology as a distinct knowledge system has opened up the possibility for such an outcome in the near future. Most historians of technology have abandoned the previous prevalent assumption that technology is only science applied for practical purposes or simply that technology is applied science.[10] Historians of science writing on the history of technology have created a variety of misconceptions about the fundamental relationship between science and technology, and their interactive relationship with society. The relationship between science and technology is dialectical, or interactive, rather than a hierarchical one in which science purportedly takes the dominant position with technology acting as science's subaltern.

The relationship between science and technology is described using various metaphors. Edwin Layton calls it "mirror-image twins."[11] George Wise describes science and technology as two autonomous spheres of knowledge, the former about the "natural world" and the latter about the "man made world."[12] Wise decries as historically inadequate the pipeline image of the science-technology relationship popularized by many historians of science, scientists and science policymakers, in which ideas first evolve within science which are then supposedly turned into useful technological products. Derek de Solla Price sees science and technology as a "pair of dancers" dancing to the "music of instrumentalities."[13] Clark regards science and technology as "distinct social systems observing different rules of conduct, coming from independent professional traditions and being largely autonomous in the influence one has on the other."[14]

The great progress in technological and scientific knowledge, particularly since the nineteenth century, clearly manifests itself in this process of mutual enrichment. In fact, given the tremendous benefits technological change has bestowed upon science, it may not be unfair to identify science as "applied technology." Though such a narrow-mindedness is undefendable, one might wonder how great advances in science would have come about without technological instrumentalities like telescopes, microscopes, electricity, mass spectrometers, nuclear magnetic resonance, and high energy particle accelerators. Science and technology are intellectually and epistemologically autonomous entities, but related to each other dialectically. The recognition of the autonomy of technological knowledge as a distinct intellectual enterprise came about as a result of the pioneering works of historians of technology. These scholars discarded the old methodology of writing the history of technology that treated technology as mere artifacts (machines and tools). Technology is being conceptualized as a body of knowledge with its own internal dynamics of change and progress. This historiographic breakthrough has made a signal contribution to the study of the change in the technological knowledge base of society. This idea will be exploited further to explain technological change in chapters eight and nine. In this chapter, however, we will examine some historically oriented models of technological change. The objective here is not to survey in a comprehensive manner how technological change is treated in the history of technology, but rather to show how it is conceptualized theoretically and how certain models of technological change with some predictive power evolved in order to apply these findings beyond the immediate context within which they first evolved. We will undertake the empirical testing of the adequacy of some selected models in chapter eight using the case study of the technological change in the Green Revolution.

CUMULATIVE SYNTHESIS OF TECHNOLOGICAL CHANGE

Presumably anticipating the intellectual trajectory of modern historians of technology, Abbott Payson Usher's "cumulative synthesis" model of technological change addresses some of the above issues of the science-technology interactive relationship.[15] The autonomy of technology is seen in this model as a distinct intellectual enterprise that is separate from its artifactual side. Usher clearly delineates the epistemic and artifactual realms of technology. Based on empirical evidence from the history of mechanical engineering, Usher emphasizes the "continuity" of technological progress, claiming that technological change is a cumulative process, in which smaller changes and inventive activities accumulate over a period of time.[16] Usher draws a sharp difference between invention and acts of skill. He claims that the "insight" required for acts of skill are within the "capacity" of any trained person. However, for inventions to occur, "the act of insight can be achieved only by superior persons under special constellation of circumstances."[17] Thus he avers, "[I]nventive acts of insight are unlearned activities that result in new organizations of prior knowledge and experience."[18]

Usher identifies three approaches to analyzing the issues involved in the inventive process: (1) the transcendentalist, (2) the mechanistic, and (3) the cumulative synthesis. He rejects the transcendentalist approach that claims that inventions are the acts of genius minds. Usher also rejects the mechanistic process that according to him is a rehashing of the old adage that necessity is the mother of all inventions. Vernon Ruttan correctly captures Usher's point that inventions are not carried out "under the stress of necessity and that the individual inventor is [not] merely an instrument of historical processes."[19] Thus Usher settles for cumulative synthesis as the right conceptual model for explaining technological change. According to Usher, the individual acts of insight which lead to an invention in the right social and economic contexts can be formalized as a "genetic sequence of four steps."[20] These steps are

(1) *perception of a problem,* which normally is an unfulfilled want; (2) *setting the stage,* which is a process that involves the experimenter (which at a lower level may be a process of trial and error and at a higher level may be a systematic process of experimentation); (3) *acts of insight,* in which the actual solution to the problem is found; and (4) *critical revision,* in which the newly perceived solutions are studied in their proper context which may lead to the development of a "technique of thought or action" similar to the evolution of an algorithm for problem solving.

Thus, for Usher, technological change is a cumulative synthesis of several small steps of inventions and improvements occurring in the life cycle of a technology. According to Ruttan, the cumulative synthesis model is appealing because it attempts to provide "a unified theory of the social processes" behind the evolution of new technologies.[21] Although Usher's social and historical explanations of the cumulative synthesis model is appealing, his fondness for reducing the inventive process to an "act of insight" within the confines of Gestalt psychology may be controversial.[22] He further claims that to have the insights for inventions, the inventor must be a "superior" person. This elitist characterization of the inventor may not always be the case as Matthew Gamser, Eric von Hippel, and Stephen Biggs show.[23] The users of technology, including ordinary consumers and farmers, are innovative and show inventive skills in helping to develop new technologies. Further, Usher's tendency to credit only the person who developed the theory or concept that enabled the invention or improvement of a technology, and to exclude the person(s) responsible for the actual invention and development of the technology in his processual model, is historically incorrect and unfair.

Usher's model of technological change follows the cumulative synthesis heuristic that emphasizes continuity in technological change. The reverse is true when one applies this processual model to case studies. That is, technologies do not become extinct in the strict Usherian sense. The "inventive" process is something in which individuals and institutions are involved, in most instances, by using concepts based on existing ideas. Although cumulative synthesis is a plausible explanation of technological change, the implicit assumption that new technologies evolve only for the purpose of functional superiority over their predecessors is not universally valid. Technological change occurs not only because of the exigencies of normative reasons like the need for functional superiority, but also due to such other factors as safety considerations, potential for labor or capital saving, facilitating automation, energy saving, and so on. In addition to these, extant disembodied factors such as social, cultural, and possibly contingent factors can also shape technological change.

SYSTEMS MODEL OF TECHNOLOGICAL CHANGE

Following Usher's cumulative synthesis model, the next most significant model of technological change to emerge in the historical tradition was the systems model developed by Thomas Hughes.[24] It is in no way implied here that Hughes was the first historian of technology to use the systems model, which Hughes readily admits in the introduction to his most important work, *Networks of Power*.[25] Hughes uses a systems framework to explain the development of a technology and its change within the context of the industrialization of three geographic regions in Western society from 1880–1930. The empirical case study that Hughes uses to support his model is the electric power system—its birth, growth, and maturity during the 50-year period. The rationale for using the systems model, according to Hughes, is that the "history of all large-scale technology—not only power systems—can be studied effectively as a history of systems."[26] Hughes shows how small inner city electric lighting systems of the 1880s grew and developed into large regional power systems of the 1920s in Western Europe and North America. He argues that it is only natural for historians to study the complexity and change, and that the electric power system is an ideal case to study and theorize about technological change.

What signifies a system is the "interconnectedness" of its components or interacting parts, and for a technological system, controls are imposed to optimize the system's performance. The goal of any system, technological, social, or sociotechnological as is the case in most instances, is to transform certain inputs into outputs under given constraints and contingencies. Therefore, the goal is to understand the system components and the overall system dynamics to stabilize and optimize its performance. Since the components are interconnected, a single component can change the system behavior and thus the system dynamics. Thus, it is of the utmost importance to find out the most efficient system configuration, the way to interconnect the system components, which are arranged either in a vertical mode or in a horizontal mode.[27] According to Hughes, those parts of the (sociotechnological) system that are not under the system's control are called the environment, a sort of backdrop. However, a component of the system can easily incorporate the environment if there is a need. A system that brings in the environment and is subject to environmental influence is called an open system.[28]

Hughes identifies five phases in the evolution of all large systems, and the dominant technological entities and professionals who shape the system characterize each phase. The invention and development of the system characterize the first phase of the model. The professionals who play the key roles in laying the foundation of the system are called inventor-entrepreneurs. They are different from ordinary inventors, because the former initiates the formation of the system at a rudimentary level.[29] According to Hughes, in the case of the electric power systems, Thomas Edison was the quintessential inventor-entrepreneur. The second phase of the systems model is the process of technology transfer from one country/region/society to another. In the case of the electric power system, a good example is the transfer of technological knowledge, the knowledge of electric transmission by stepping up and down voltages to facilitate long distance transmission of power from USA (New York City) to Germany (Berlin) and Britain (London). A key technological invention that facilitated this knowledge transfer was the transformer that utilized the principle of in-duction. Hughes argues that a variety of factors, such as the nature of the technology, politics, and legislative prerogatives, influence the course and process of technology transfer. The agents of technology transfer and the associated system leaders include inventors, entrepreneurs, financiers, and engineers.

The third phase of the model is system growth. The particular intellectual tools that Hughes invokes to analyze system growth are "reverse salients"[30] and "critical problems." Technological imbalance or reverse salients evolve due to uneven development of certain system components that become critical problems impeding the smooth growth of the system. During this formative stage in the system development, engineers, inventors, and other professionals consolidate their "constructive powers" to change reverse salients in order to steer the system away from bottlenecks that can frustrate the attainment of system goals. As the system builders come across reverse salients, they identify and later analyze them as a series of "critical problems" to be solved as a means to save the system from collapsing. Therefore, "[D]efining reverse salients as critical problems is the essence of the creative process, . . . [and] when engineers correct reverse salients by solving critical problems, the system usually grows if there is adequate demand for its products."[31] In the case of the electric power system, an earlier reverse salient was the difficulty of transmitting direct current to consumers far away from the generating stations.[32] The invention of first the thyrister and later the synchronous alternating current generators, and ultimately electric transformers, solved this major limitation of direct current.[33]

The fourth phase of the systems model is "momentum," and as is the case with all systems having substantial momentum, it acquires "mass," "velocity," and "direction" during the growth phase. The mass in a technological system consists of machines, physical structures, and other material artifacts that manifest in the form of the considerable capital investment made by the system builders. For electric utilities, the mass consists of generators, steam turbines, transformers, substations, power lines, distribution networks, and other power system components. A system in growth phase acquires a system culture that is characterized by the involvement of skilled engineers, government regulatory agencies, business concerns, educational institutions, professional societies, and other organizations that shape the technological and business "core" of the system. According to Hughes, a system with such momentum gathers a perceptible rate of growth or velocity.[34] The system acquires a goal or direction as the system accelerates to attain a steady and self-sustaining growth. The goal of the electric utility system then turns into a race to utilize its natural monopoly powers in a captive market.

The last phase of the system model is characterized by a qualitative change in the nature of the system's stability and its problem-solving culture. During this phase, reverse salients emanating from technological problems give way to reverse salients caused by problems of financial matters and management of large regional systems with substantial capital investment. During this mature phase of the system, the leadership of the system is in the hands of consulting engineers, utility managers, and financiers.[35] The primacy of the inventor-entrepreneurs who laid the foundation of the system attenuates during this stage. Consulting engineers and professional managers dominate the management of the system during this stage.[36]

In a valiant effort to broaden his systems model to include relevant findings from other disciplinary investigations like economics and sociology, particularly the social constructivist interpretations of technology espoused by the latter, Hughes makes the claim that all technological systems are social constructions that are made out of a "seamless web" of science, economics, and social forces. Hughes's claim is that since technological systems are "invented and developed by system builders and their associates, the components of technological systems are socially constructed artifacts."[37] Hughes, however, differs from the actor-network theory of Michel Callon and others (see the next chapter) in such a way that he includes only the animate (reflexive) actors, such as inventors, industrial scientists, engineers, financiers, and managers, in his network. Despite this, Hughes claims that his systems model of technological change and the actor-network and social constructivist models have several things in common. In the five stages described earlier, according to Hughes, one may find all the standard structural components of technological change—invention, development, innovation, transfer, diffusion, growth, competition, and consolidation—though not necessarily in that order.[38]

Hughes enlarges the scope of the systems model further by incorporating the above structural components of technological change. Outstanding examples of independent inventors and their radical inventions that sowed the seeds of large technological systems are Bell and the telephone, Edison and electric power and light, the Wright Brothers and the airplane, Marconi and the wireless and radio, and Elmer Sperry and the gyrocompass guidance and control system (of airplanes, ships, missiles, and so on). During the invention, innovation, and development phase of the system, the key actors are the inventor-entrepreneurs. Hughes claims that the inventors who lay the foundations of new systems are usually outsiders. Existing firms and companies that evolved from the rudimentary systems ventures do not like the radical inventors who founded the systems. That is why radical inventors start their own technological enterprises, most of which eventually become large technological systems. During the development phase, "inventor-entrepreneurs and their associates embody in their invention [all the] economic, political, and social characteristics that it needs for survival in the use world."[39] During the innovation phase, the inventor-entrepreneurs along with the engineers, industrial scientists, and others help to bring the product into use. During the transfer, diffusion, growth, and competition phases, the key actors are manager-entrepreneurs. Finally, financier-entrepreneurs and consulting engineers dominate the consolidation and rationalization phase of the system. While the transfer phase displays distinct styles of system performance, the growth, competition, and consolidation phases show the momentum of the system. After reaching momentum, most systems continue in that state for a long time and eventually reach a stage of "stasis."[40] Terry Reynolds, though highly appreciative of Hughes's meticulous research and scholarship, nevertheless raises the general concern whether "system" has applicability as a unit of analysis of Hughes's particular case study, or whether it can be used as a standard unit for all cases of technological change.[41] John Law questions Hughes's indifference on using the same analytical method to explain micro and macro phenomena.[42] Despite these concerns, Hughes's model of technological change is comprehensive and very useful to study the phenomenon of how technology evolves and changes in industrializing societies.

TECHNOLOGICAL PARADIGMS AND REVOLUTIONS

One of the most successful applications of Kuhn's model of scientific change to explain technological change in the historical tradition has been attempted by Edward Constant.[43] Using the technological change in aircraft engine propulsion systems, which reflected the change from piston-engine driven propellers to gas turbine driven jet engines, as empirical support, Constant claims that technological change and progress occur analogously (with some variations added) as scientific change expounded by Thomas Kuhn.[44] According to Constant, the turbojet revolution is exemplified by the paradigmatic shift from propeller driven aircraft to jet propelled aircraft during the 1940s. A technological paradigm, for Constant, is an "accepted mode of technical operation, the usual means of accomplishing a technical task."[45] Constant bases his model on a claim that "technology is intrinsically imperfect," albeit the possibility that it has almost infinite capacity for becoming better, faster, safer, and more efficient, unlike science, which is "capable of attaining a perfect match between observation and theory."[46] Although Constant rightfully points out that technological revolutions normally occur because of functional failures, in his model, however, technological revolution occurs due to what he calls "presumptive anomaly."

Presumptive anomaly is said to occur when scientific insights or assumptions that are derived from the latest development in a specific technology's cognate scientific domain indicates "either that under some future conditions the conventional system will fail (or function badly) or that a radically different paradigm will do a much better job or will do something entirely novel."[47] Thus, Constant gives science the sole causal agency for initiating paradigm change in technological practice. The existing technology works perfectly well under present functional parameters and expectations. However, an anomaly would occur if the functional parameters were to change, which could be predicted because of advances in the science of aeronautics in the case of airplanes, or whatever scientific principles that form the theoretical backbone of the purported technological enterprise. Presumptive anomaly is thus touted as the critical link between science and technology.[48] The presumptive anomaly, in turn, paves the way for one or more "candidate" paradigm(s). When the winning paradigm can conclusively show the comparative functional failure of the conventional technological system and its practice, it garners the loyalty of a majority of technology practitioners.

The turbojet revolution took place at a time when the conventional piston-engine driven propeller planes were enjoying spectacular success. No functional failure had occurred, none was in sight, and the conventional propulsion system was exploiting advances in airframe designs, accessories, and the latest aircraft manufacturing technology. The presumptive anomaly was noticed by a handful of practitioners who were aware of advances in aerodynamics—the study of gas-flow behavior. During the 1920s, while investigating the flight performance under transonic conditions some practitioners of the aircraft design and manufacturing community observed the phenomenon of propeller blade-tip compressibility burbling. As the speed of aircraft went up to sonic levels, the blade tip compressibility burbling could cause flight instability and fateful accidents. That is, in the future, the existing system would fail if a key operational parameter—speed—were increased. It was also revealed from the study of B. M. Jones in 1927 that current aircraft wasted at least two-thirds of their power in overcoming unnecessary turbulence while in flight. Given spectacular advances in other areas of aircraft design and operation, propeller streamlining became an accepted norm for increasing speed and efficiency. But the theoretical knowledge of blade-tip compressibility burbling could thwart these efforts because of the presumed anomaly. Thus, some practitioners (a prominent candidate was Frank Whittle, who developed the first jet engine for airplanes) put forward a candidate paradigm of a radically new propulsion system using gas compressors and turbines that could do away with propellers. Only a few practitioners or "provocateurs" (who are mostly "outsiders"—not regular members of the technological community) foresaw the superior efficiency of turbojet engines,[49] while the traditional practitioners stuck with the existing piston-engine driven propulsion system. However, eventually the turbojet propulsion system overthrew the "normal" aircraft (piston-engine) propulsion system and installed a new technological paradigm (turbojet engine).

Constant admits the importance of nonscientific factors such as economics, patents, and cultural factors like values and norms in the formation and acceptance of a new technological paradigm. However, he argues that economic factors may play a more deterministic role in the process of diffusion of the new technology, especially such factors as firm-by-firm adoption of the new innovation.[50] Constant correctly points out that patent disclosures cannot be a significant factor in the case of the technological change in aircraft technology, as the number of patents were very few, and many of the key players who spearheaded the technological revolution "developed their ideas in ignorance both of prior patents and of each other, and [these] were successful despite what turned out to be conflicting patent claims."[51] Despite admitting the significance of values and belief system in shaping technological change, Constant does not delve into these important issues in his model.[52] Constant's claim that it is outsiders, who most often are not regular members of the community of technological practitioners, who perceive presumptive anomaly and set off technological revolutions should not be taken as a universally valid axiom. This may be just a contingent factor that is particularly amenable to the case of the turbojet revolution. Despite the arrival of the jet aircraft, propeller planes did not disappear from the sky. Both coexist today.

The "community of practitioners" forms the formal unit of analysis of Constant's model of technological change, where "technological knowledge is expressed in well-winnowed traditions of practice that are the possession of well-defined communities of technology practitioners."[53] David Lewis and Edward Ezell find Constant's work highly evocative and claim that this model is exemplary without offering any reason why it is so.[54] This enthusiasm has to be qualified because it is not necessary that there should be a presumptive anomaly in the history of a technology for its replacement by a better one. Also, Constant's claim that the emergence of revolutionary ideas, or new ways of conceptualizing alternative technological solutions, always come from the outside of the community of technology practitioners was a point already raised by others. This claim, irrespective of authorship and priority claims, is not historically correct.[55]

CONTINUITY AND EVOLUTION IN TECHNOLOGICAL CHANGE

Cumulative change as a historiographic tool to analyze the development of technology was a powerful intellectual motif that historians of technology took to their hearts. A. P. Usher's "cumulative synthesis" model of technological change we discussed above is a variant of this concept. Darwin's and other evolutionists' hypothesis of natural history stated that new species evolve through a process of natural selection. This hypothesis gave historians of technology a new historiographic idea to reformulate evolutionary change into a potentially powerful heuristic to explain technological change. George Basalla attempted one of the boldest reconstructions of technological change as an evolutionary process in the history of technology.[56] Basalla argues that in understanding technological change, the artifact takes the central conceptual role, and hence, the artifact forms the primary unit of analysis of his model. Basalla correctly points out that technology predates science as long as human civilizations have existed. He sees the end product of science or the unit of analysis of science as "written statement" and the corresponding end product of technological activity, an "addition to the made world," or simply an artifact.[57] One of the major theoretical assumptions of Basalla is that, unlike scientific change, which is characterized by discrete events (revolutions), technological change is a continuous and cumulative process that is characterized by evolutionary change.[58]

Basalla presents numerous cases of technological artifacts such as the cotton gin, steam and internal combustion engines, electric motor, transistor, incandescent lamps, and barbed wire, among others, to explore the evolutionary nature of technological change. He disputes the heroic inventor model of technological development as being ahistorical. For example, it is popularly believed that Eli Whitney invented the cotton gin while touring the southern United States where he saw slaves separating cotton fibers from the seeds. Long before Whitney invented the cotton gin, mechanical cotton gins were in wide use in the South. The mechanical cotton gin was a variation of the Indian *charka* that was in use in India before the Christian era.[59] This technological device was available to the Italians during the twelfth century, although it is not clear how this device reached Italy. It was mentioned in Chinese writings of the fourteenth century, and in Diderot's *Encyclopedie*. In 1725, the roller gin (the most updated variation of the *charka*) was introduced into the Louisiana Territory from the Levant.[60] Eli Whitney first encountered the roller gin in the cotton growing South in 1793.[61] Highlighting the continuity theme of evolutionary change in technological development, Basalla thus concludes, "[A]cknowledging the *charka,* however, shows that Eli Whitney's invention had artifactual antecedents whose overall structure and mechanical elements were adapted by the American inventor to suit his purpose."[62]

The transistor is another technological device that Basalla tries to save from the grist of the revolutionary model mill of technological change. Basalla shows that the transistor can be traced to a series of scientific works in the nineteenth century. The step leading to the development of the transistor can be traced to the work of Ferdinand Braun, who in 1870 discovered that certain crystalline substances conducted electricity only in one direction. At the beginning of the twentieth century these crystal rectifiers were used to detect radio signals. Though the crystal radio signal receiver was a semiconductor device, it failed as a radio set because it could not amplify the received signals. However, the invention of the vacuum diode and later the triode made modern radio possible.[63] Basalla claims that the development of the modern transistor was not a revolutionary replacement of the vacuum tubes, but a continuation of the semiconductor work of Braun from the nineteenth century to the crystal detectors of the early twentieth century.[64] The inventors of the transistor, John Bardeen, Walter Brittain, and William Shockley, were working on an existing technological device to solve an intractable switching problem of long distance telephone lines of their employer, the American Telegraph & Telephone (AT&T). One of the key elements in a successful telephone network is the switching device in the exchanges. The electromechanical switches used vacuum tubes and mechanical relays to complete the connection between subscribers. However, because of the rapid wearing of the mechanical parts and the unreliability of the vacuum tubes due to overheating, electromechanical relays became a major reverse salient in the growth of telephony, until transistors replaced the cumbersome vacuum tubes as switches. The transistors eventually led to the development of fully electronic switches, without any moving parts. The advent of digital technology is now revolutionizing telecommunication systems.

Another example, related to Basalla's model, where temporal factors do not mesh with the historical perception is the so-called invention of the steam engine by James Watt. Harnessing the latent energy of falling water to provide power for mechanical devices was known to many cultures since antiquity. The conversion of the stored energy in combustible substances like coal and wood into steam power to run pumps, looms, mills, and other machines was a giant step that greatly helped the British Industrial Revolution.[65] The precursor of Watt's steam engine was the Newcomen engine that was in operation to pump water out of flooded mines in England. In the early versions of the Newcomen engine, the piston was pulled to the top of a vertically mounted cylinder by the weight of a lever arm. While the piston moved upward through the cylinder, steam was allowed to enter the cylinder from below. While the piston would rise to the top of the cylinder and the cylinder would be filled with steam, the cylinder was cooled with cold water. As the cylinder was cooled, the condensing of the steam inside the cylinder would cause a partial vacuum. Consequently, the atmospheric pressure acting on the piston would force it to move downwards, simultaneously filling it with air. Thus, technically, the Newcomen engine was actually an atmospheric engine. What James Watt did in 1769 was to make the Newcomen engine a more efficient machine by adding a separate condenser, instead of cooling the cylinder for every return stroke. Watt's engine subsequently became a "double-acting reciprocating engine." Watt's engine became useful because of its portability and adaptability to rotary motion, and thus became useful for technological devices like electrical generators, locomotives, ships, and numerous other industrial machines. Eventually, the internal combustion engine became possible because of the functional inadequacies and low efficiency of the steam engine.[66]

Although Basalla strongly defends the continuity argument, he refrains from the claim that inventions are inevitable. In order for the new invention to appear based on an existing artifact, the right social, cultural, economic, and technical forces are needed as important agents of selection leading to the new artifact. What matters are the understanding and the ability to search for "novelty" by the inventor. In order for technological change to occur in an evolutionary fashion, "then novelty must find a way to assert itself in the midst of the continuous."[67] In the search for novelty, psychological and intellectual factors take precedence over social and economic conditions. Finally, Basalla demystifies any notion of technological progress. He resists the "tendency to make the advancement of humanity or biological necessity the end toward which all technological change is directed."[68] Basalla's claim that technological advance does not have a causal connection to the betterment of humanity, however, is an argument that can hardly be supported. However, whether technological advances made modern human beings better people with social grace and conscience than their ancient counterparts is a moot point that is loaded with normative and metaphysical connotations.

John Smith, another strong defender of evolutionary change, claims that, in order to capture the complexity of technological change, particularly in its developmental phase, historians need to use evolutionary theory.[69] He argues that much of historical work on technological change followed a linear model of accounting for the sequence of invention, innovation, and development. In the greater scheme of things, invention took center stage, because without it, the latter cannot be there. He was correct in pointing out that the linear model is simplistic, ahistorical, and captures only successful innovations. We know that only a small fraction of innovations ever end up as technologically useful products and services. Smith claims that the key to capturing the complexity of technological change through the evolutionary model is to concentrate on the developmental process. If the particular artifact cannot find a "niche in the larger environment," it will not be a successful innovation.[70] Generation of novelties and niche finding are the two conceptual issues one needs to locate in explaining the innovation process. According to Smith "[I]nnovation occurs when certain technological novelties are *selected* from the wide array of choices and put into a niche in the larger technological environment."[71]

Smith uses the development of celluloid plastic as a case to illustrate his "nonlinear" model. The story starts with the invention of a substitute material for ivory, which was in short supply during the late nineteenth century in the United States. This contingency led to the price of ivory products such as billiard balls going sky high. Innovators noticed that a potential market existed for a substitute for ivory. An Albany, New York, inventor by the name of John Hyatt experimented with cellulose, which was discovered by chemists a few decades earlier. Hyatt found out that adding camphor to nitrated cellulose yielded a compound that was malleable and stable. The billiard balls made out of this new plastic were, however, rejected by the players who preferred the expensive and hard to find ivory balls. If we used the linear model of technological change, the story ends there. The new product was rejected by the consumers and failed to capture a niche in the larger environment. Undaunted, Hyatt looked elsewhere. He found his new compound useful as a material for combs, brushes, and mirrors (products that were previously made of ivory), which had both elements of novelty and the right niche. Nevertheless, the brightest future for Hyatt's celluloid plastic was found in the form of films in still photography and motion pictures. The complexity of the developmental process of celluloid plastic, therefore, is captured best by an evolutionary paradigm rather than by a simplistic linear paradigm.

TECHNOLOGY AS KNOWLEDGE AND TECHNOLOGICAL CHANGE

The most original theoretical contribution to explaining technological change by historians of technology was the bold historiographic methodology of reconstructing "technology as knowledge." The theoretical conundrum of how to account for the growth of knowledge in the history of technology became a serious problem for those interested in not seeing technology as the handmaiden of science. The question of what exactly is the subject matter of the history of technology became a serious issue when technology was found to be intellectually autonomous from science. Is it a chronicle of the evolution of technological artifacts? Is it an interpretation of the artifacts themselves without any normative content? How and where should the agency question be addressed while chronicling the progressive evolution of technologies? Can reflexivity be attributed to the artifacts themselves in explaining how technological change encompassed a progressive element in the narrative scheme? These are serious issues for the new field that had weaned itself from the imperious history of science parent.[72]

The ahistorical suggestion that technology is merely applied science is an epiphenomenon of the enormous success of science as pre-eminent knowledge. As de Solla Price pointed out earlier, the characterization of technology as only an application of science for material ends was the creation of shoddy historians of science and science policymakers.[73] What caused science to become ubiquitous with knowledge was merely a contingent development. So equating technology with science was no more logically implausible, as historians of technology recognized that what they were interpreting was not the physical reconstruction of artifacts but the intellectual and cognitive aspects of the process of technological change itself. Although the intellectual autonomy of technology as knowledge has been affirmed, science does provide a significant fraction of the knowledge spectrum for technology. It evolves out of the interactive or dialectical relationship between science and technology where mutual enrichment of knowledge content is the norm.[74]

Edwin Layton and Walter Vincenti see epistemological issues as prominent in technology because of the problem of finding out what engineers do and how they account for progress and change in their respective disciplines.[75] Vincenti and Layton use "design" as the unit of analysis to trace the growth and change of engineering knowledge qua technological knowledge. According to Layton: "[A] design embodies the knowledge needed to produce a technological device or a system. It constitutes the cognitive bridge across a spectrum from abstract, idealized conceptions to the concrete, highly complex products of technology existing in the real world."[76] Using design as the metaphor for engineering knowledge solves two major problems. One is the actual physical configuration of the artifact and the second one is the cognitive nature of engineering practice. A change in design reflects the knowledge change and the artifactual change. Design is a social activity "directed at a practical set of goals intended to serve human beings in some direct way."[77] John Staudenmaier theorizes technological change as manifested by a "tension between technical design and its ambience."[78] The "ambience" could be collectively characterized as the sociocultural context or the selection environment.

Walter Vincenti traces the growth of knowledge and the ensuing technological change through the empirical example of airplane design changes, particularly changes in airfoil design.[79] Through the exploration of the design process at Consolidated, an aircraft maker based in San Diego, California, during the early part of this century, Vincenti shows how "an engineering design community functions in the face of uncertainty at a given period of time."[80] The uncertainty in design and functional parameters was further vitiated by the unexpected airfoil design by an outsider (outsider to the community of aircraft designers) named David Davis in the 1930s. The larger design community was skeptical of Davis's airfoil for lack of any sophisticated theoretical analysis of fluid dynamics. Nevertheless, one major company (Consolidated) adopted Davis's airfoil for long-distance bomber aircraft.

However, subsequent advances in aerodynamics theory could not verify the superior performance standards claimed by Davis. It was noted that when aircraft speed increased, Davis's airfoil would encounter functional failure analogous to the presumptive anomaly concept developed by Edward Constant.[81] However, in the uncertain period of the 1930s, Davis's airfoil enjoyed a measure of acceptance.[82] It can be argued that despite its eventual rejection by the aircraft designers, Davis's airfoil did play a role in the growth of knowledge and progress in the practice of engineering. Davis's airfoil may be only a "footnote" in the annals of aeronautics history and appear to have wasted much time for the community of engineering practitioners, but it is precisely these mundane and unimportant trial-and-error episodes that make engineering practice a human endeavor. Without these activities, it is hard to achieve technological change and progress. Vincenti presents the nonselection of the Davis airfoil as an example of the "blind-variation-selection" process of technological change, an important concept that will be taken up in chapter five. The concept of technology as knowledge is further developed in chapters seven and nine by following a different conceptual path from Vincenti's and Layton's. As promised at the beginning, let us return to the theoretical focus and emplotments of metahistorians to explain technological change.

METAHISTORY AND METASOCIOLOGY AS FRAMEWORKS FOR TECHNOLOGICAL CHANGE

As mentioned earlier, despite the paucity of their application (or perhaps their difficulty in adapting to empirical case studies), the modes of narration of metahistorians and metasociologists can nevertheless be developed into useful theories of technological change. The narrative modes of one metasociologist—Max Weber—and three metahistorians—Oswald Spengler, Karl Marx, and Arnold Toynbee (who are identified as "speculative philosophers of history" by cultural critic and philosopher of history Hayden White[83])—are presented below in order to tease out, from their specific sociological and historical modes of narration, useful theories of technological change.

Max Weber's sociohistorical works have been, largely, to expound the dynamics of Western civilization and the rationality of its intellectual and economic foundations.[84] In addition to the Protestant ethic that underpins the metaphysical foundation of modern capitalism that originated in the West, Weber credits "science and scientifically oriented technology" for the rational and intellectual articulation of capitalism. According to Weber, the ontological foundation of Western civilization can be found in its proclivity for "intellectualist rationalization," which is articulated through science and industry. Mikael Hård uses the Weberian notion of "intellectuallist rationalization" to explicate the technological change that had taken place in refrigeration and brewing technology in nineteenth-century Europe.[85] Hård sees a parallel between Weber's theory of rationalization and the historical development of the West to the scientification of the technology of mechanical refrigeration. Hård shows how the German scientist Carl Linde translates the science of thermodynamics and Linde's ontological predisposition into the construction of a workable ice machine. Hård argues that the scientification of refrigeration took place between the period 1870–1893 during which time Linde published seminal papers on the application of the theories of thermodynamics to refrigeration. Further progress in refrigeration technology occurred as a result of the enthusiastic reception of the ice machine in the brewing industry. This is an excellent case study of a "Weberian approach to the history and sociology of technology" that clearly explicates the claim that all social phenomena, including technological change, "are constituted through a process of conflict."[86]

In many ways similar to historical sociologists, the role of historians is to sort out the various historical "facts" and evidence available to him and interpret these to collate and create a narrative account of historical events following a particular theoretical framework or narrative mode. According to White, a historical narrative is "a mixture of adequately and inadequately explained events, a congeries of established and inferred facts, at once a representation that is an interpretation and an interpretation that passes for an explanation of the whole process mirrored in the narrative."[87] As Kuhn, White, and other historicists have shown, there is no theory-neutral way of observing and then theorizing and interpreting social or scientific or historical phenomena. Darwin was the quintessential scientist whose scientific practice showed the theory-ladenness of (scientific) observations.[88] Models of technological change may be constructed very much in the spirit of the scientific practice of Darwin (and other great scientific practitioners) and the narrative modes adopted by the metahistorians. Thus, the objective here is to extrapolate from the theoretical or narrative modes of Spengler, Marx, and Toynbee to apply these extrapolations to explicate technological change with all the appropriate caveats in place. These concerns include the usefulness of these metatheoretical frames, particularly in the wake of the postmodernist angst about metanarratives and metatheories underrepresenting social realities and historical verities.

In *The Decline of the West,* Oswald Spengler postulated a life cycle theory of the birth, growth, and decline of all human cultures and civilizations. According to Spengler, civilization is the "organic destiny" of all cultures, and "every Culture has *its own* Civilization."[89] He argued that, despite their morphological differences and diverging value systems, from a comparative morphology of world history perspective, all cultures exhibited the above-mentioned stages in their transition from culture to civilization. All coherent and successful cultures are held together by a set of fundamental value systems which guide the members of each culture. These values give the members their identity and provide normative parameters to construct themselves as members of particular communities and societies. These fundamental values and beliefs can be identified as the worldview or "axiomatic core" of each culture.[90] These worldviews act as a sort of heuristic or internalized algorithm to construct the art, literature, science, religion, social relations and institutions, and coherent knowledge systems during the life course of each culture. As long as the axiomatic core provides the spiritual and intellectual sustenance to each society, it sustains and nurtures all subsequent generations. When the worldviews come under attack and cannot hold the societies together, the culture enters a period of stasis and eventually declines if it cannot find another worldview to replace the failed one.

Spengler's narrative mode of world history can be applied to explicate the process of technological change. The construction of every technology is based on a distinct axiomatic core that is made up of specific scientific and engineering principles, and a clear set of functional heuristic and operational principles. As long as these basic principles are elaborated and articulated in a coherent manner, the technology develops and further improvements are anticipated. However, technological change reaches a stasis once the axiomatic core cannot sustain growth and provide new ideas for improvement of the technology. A radically new technology is the only solution likely to overcome the stasis. The case of technological change in computers in which the analog system has given way to a revolutionary digital system is a good example. The inherent limitations of computing speed, construction of logic systems, materials, size, and operating systems of analog computers reached a stasis. It was then discovered that a radically new concept of building logical circuits using the principle of Boolean algebra was possible. The further articulation of the digital logic system, from vacuum tubes to transistors, and from transistors to integrated circuits revolutionized computers. Advances in semiconductor technology and new operating systems have taken digital computers to greater levels of speed and versatility in their applications to human endeavor. However, the limits of semiconductor devices as switches will lead to optical computers where electrons as carriers of digitized messages will be replaced by photons as the carrier signal for logical switching operations. The endless formation, transformation and decline of technologies, following a spatio-temporally circumscribed Spenglerian life-cycle model of technological change, with a unique axiomatic core, is self-evident in the above case study of computers.

By inverting the dialectics of Hegel and the materialism of Feurbach, Karl Marx formalized his theory of dialectical materialism.[91] From Marx's perspective, historical change is seen as a series of progressive evolutionary changes (but revolutionary transitions) from slave system to primitive communism to feudalism to capitalism, and finally to communism through the transitional stage of socialism, at which point human history is supposed to end. To some extent, the heuristic inherent in the idea that the contradictions in the material relations and the mode of production in each social system leads to its unravelling and eventual overthrow by a rationally superior system can be applied to explain technological change. The development of better technologies as a resolution of the inner contradictions in the nature and logical configurations of the previous technologies forms the heuristic of this model of technological change. The transition from hand loom to power loom can be explained as the selection of a functionally superior technology. The new selection environment, operating at the production level, supposedly occurred due to the adverse relations of production that existed in the capitalist economy in which the hand-loom worker lost the autonomy and power over his means of production. The application of this model to a more recent case that captures the "internalist" account of technology's development is the development of cleaner and safer electric power generation systems that evolved as a result of the contradictory social relations created by such negative externalities as pollution and safety hazards to workers and citizens. The resolution of this contradiction called for radically new technologies that would limit or avoid these negative externalities. The technological change in electric power systems can be partly explained, independently, as a result of the adverse environmental, economic, and other problems associated with the system as explained by Hughes as we saw earlier.

Like Spengler, Arnold Toynbee in his *A Study of History*[92] set out on a valiant quest to "understand the mystery of the rise and fall of civilizations."[93] Toynbee's narrative mode can be distilled as a systematic approach to explain the "genesis, growth, breakdown, and disintegration" of the "species of society" known as civilizations.[94] Toynbee identified religion as the core of all civilizations. The narrative mode of Toynbee that can be adapted for understanding technological change is his emplotment mode of "challenge and response," his attempt to explain how civilizations began. This attempt also includes such issues as how civilizations faced challenges from within and outside, how they tried to resolve these challenges, or how they disintegrated and/or are in the process of disintegration. Thus, technological change can be interpreted as the way that technologies evolve and change through the agency of individuals and institutions in response to the challenges proposed by the functional requirements, environmental conditions, markets, competitors, and contingencies. The development of catalytic converters for automobiles or scrubbers on smokestacks of thermal power plants and steel plants, among others, can be analyzed as contingent upon environmental regulations imposed by the government on electric utilities to reduce pollution. These technological developments came about as a result of the challenges created by the regulators, consumers, and the market forces.

Although no scholar has attempted to use these theoretical frames and emplotments to model technological change, their heuristic power is limited to explain this phenomena because of their inability to delve into the micro realities of technological change. However, with appropriate modifications and variations on the theme, these frameworks may show some potential.

CONCLUSION

Theories and models of technological change in the historical tradition can provide a variety of ways to analyze and understand this complex phenomenon. There is no unified approach to explaining technological change in the historical tradition. It varies from such macro approaches as metatheoretical frames, revolutionary change (paradigm shift), and evolutionary change and systems, to micro approaches as cumulative synthesis and technology as knowledge. Except for those adhering to the revolutionary explanation of technological change, all other models here take evolutionary change operating through a general selection mechanism or background environment to be, more or less, an unchallenged norm. The background environment could be as specific as particular markets and communities of practitioners and as general as institutions, cultures, markets, and social systems.

ENDNOTES

1. Edward W. Constant, *The Origins of the Turbojet Revolution* (Baltimore: Johns Hopkins, 1980), 2.
2. As will be shown in a later chapter on this topic, Mikael Hård, *Machines are Frozen Spirit: The Scientification of Refrigeration in the 19th Century: A Weberian Perspective* (Boulder, CO: Westview Press, and Frankfurt: Campus Verlag, 1994), developed a very interesting Weberian interpretation of the history of refrigeration and brewing technology in the nineteenth century. Also, Louis A. Girifalco, *Dynamics of Technological Change* (New York: Van Nostrand, 1991), provided a short narrative account of how some of these metahistorians' theories of historical change may be applied to understand technological change. However, he does not develop these theories for further elaboration because his focus was not on developing a historical model of technological change. In fact, he offered an engineering systems model to explain technological change.
3. Nathan Rosenberg, *Inside the Black Box: Technology and Economics* (Cambridge: Cambridge University Press, 1982), 34.
4. Economic historians like Nathan Rosenberg, *Inside the Black Box,* and sociologists like Donald MacKenzie, "Marx and the Machine," *Technology and Culture* 25 (1984): 473–502, were among those who ventured to write about Marx's bold explorations into the history of technology.
5. Karl Marx, *Capital: A Critique of Political Economy,* Volume I, tr. Ben Fowkes (New York: Vintage Books, 1977), 493.
6. My attention to this observation of Marx in a footnote to his Volume I of *Capital* was drawn by Rosenberg, *Inside the Black Box,* 34.
7. Rosenberg, *Inside the Black Box,* 35.
8. I do not suggest that there is unanimity among historians of technology to follow a unified methodology of either internalism or contextualism. What is being implied is that historians of technology do treat technology as intellectually distinct from science. For an interesting discussion of the methodological and conceptual debate in the discipline of history of technology, see Robert Fox, "Introduction: Methods and Themes in the History of Technology," in *Technological Change: Methods and Themes in the History of Technology,* Robert Fox, ed. (Amsterdam: Harwood Academic Publishers, 1996), 1–15.
9. I do not claim that Kuhn's model of scientific change was universally accepted as the only one. There are numerous criticisms of the Kuhnian model. For more details on this score, see endnote 3 in chapter one.
10. Derek de Solla Price, "On the Historiographic Revolution in the History of Tech-

nology: Comments on the Papers by Multhauf, Ferguson, and Layton," *Technology and Culture* 15, no. 1 (1974): 42–48, and "Is Technology Historically Independent of Science? A Study in Statistical Historiography," *Technology and Culture,* 6, no. 4 (1965): 553–568, argues forcefully that some science historians and science policymakers popularized this misperception. A few other historians of technology who uphold the epistemic autonomy of technology are: George Wise, "Science and Technology," *OSIRIS,* 2nd Series, 1 (1985): 229–246, and John J. Beer, "The Historical Relations of Science and Technology: Introduction," *Technology and Culture* 6, no. 4 (1965): 547–552.

11. Edwin T. Layton, "Mirror-Image Twins: The Communities of Science and Technology in the 19th-Century America," *Technology and Culture* 12 (1971): 562–580.

12. Wise, "Science and Technology."

13. Price, "Historically Independent." Price attributes to Arnold Toynbee the dancing pair metaphor for science and technology. Toynbee, *A Study of History,* revised and abridged (Oxford: Oxford University Press, 1962), attributes the relationship between "Physical Science" and "Industrialism" as a pair of dancers.

14. Norman Clark, "Similarities and Differences Between Scientific and Technological Paradigms," *Futures* 19, no. 1 (1987): 26–42, 26.

15. Abbott P. Usher, *A History of Mechanical Inventions* (Cambridge, MA: Harvard University Press, 1954), "Technical Change and Capital Formation," in *The Economics of Technological Change: Selected Readings,* Nathan Rosenberg, ed. (Harmondsworth, UK: Penguin Books, 1971), 43–72.

16. This view of Usher's may be contrasted to his colleague Schumpeter's to understand the dichotomy in their views on invention and innovation. Schumpeter argues that "innovations are changes in production functions which cannot be decomposed into infinitesimal steps. Add as many mail-coaches as you please, you will never get a railroad by so doing." See Joseph Schumpeter, "The Analysis of Economic Change," in *Essays of J. A. Schumpeter,* R.V. Clemence, ed. (Cambridge, MA: Addison-Wesley, 1951), 134–142, 136. For more on the difference between Usher and Schumpeter on technological change, see Govindan Parayil, "Schumpeter on Invention, Innovation, and Technological Change," *Journal of the History of Economic Thought* 13 (1991): 78–89.

17. Usher, "Technical Change," 43–44. He contended that Gestalt Psychology could explain these differences. According to Usher, *Mechanical Inventions,* 61, "[T]he Gestalt analysis presents the achievements of great men as a special class of acts of insight, which involves synthesis of many items derived from other acts of insight."

18. Usher, "Technical Change," 46.

19. Vernon W. Ruttan, "Usher and Schumpeter on Invention, Innovation and Technological Change," in *The Economics of Technological Change: Selected Readings,* Nathan Rosenberg, ed. (Harmondsworth, UK: Penguin Books, 1971), 73–85, 78.

20. Usher, *Mechanical Inventions,* 65.

21. Ruttan, "Usher and Schumpeter," 80.

22. This argument, notwithstanding Usher's, *Mechanical Inventions,* 61, caveat, that "Gestalt psychology affords an explanation of the process of invention that is intermediate between the mystic determinism of the transcendentalists and the mechanistic determinism of the sociologic [sic] theories." Usher, *Mechanical Inventions,* 83, further adds that "the completion of the analysis of the processes of innovation will necessarily be the work of the psychologists."

23. Matthew S. Gamser, "Innovation, Technical Assistance, and Development: The

Importance of Technology Users," *World Development* 16, no. 8 (1988): 711–720; Eric von Hippel, *The Sources of Innovation* (New York: Oxford University Press, 1988); and Stephen D. Biggs, "Informal R&D," *Ceres* 13, no. 4 (1986): 23–26.

24. See Thomas P. Hughes, *The Networks of Power: Electrification in Western Society, 1880–1930* (Baltimore: Johns Hopkins, 1983), "The Evolution of Large Technological Systems," in *The Social Construction of Technological Systems: New Directions in the Sociology and History of Technology,* Wiebe E. Bijker, Thomas P. Hughes, and Trevor Pinch, eds. (Cambridge, MA: MIT Press, 1987), 51–82, "The Seamless Web: Technology, Science, Etcetera, Etcetera," *Social Studies of Science* 16 (1986): 281–292, and "The Development Phase of Technological Change," *Technology and Culture* 17, no. 3, (1976): 423–431.

25. Although originally conceived for solving engineering problems, systems as an organizing concept or model have permeated most disciplines and areas of intellectual inquiry, like history, sociology, management, biology, and economics. In sociology, for example, Talcott Parsons, *The Social System* (New York: Free Press, 1964), used the frame of systems for the analysis of the structure and processes of social systems. For Parsons, the frame of reference for social analysis is the actions and interactions of individuals. Returning to the history of technology, French historian of technology Bertrand Gille, "Prolegomena to a History of Techniques," in *The History of Techniques,* Volume I, *Techniques and Civilizations,* Bertrand Gille, ed. (New York: Gordon and Breach Science Publications, 1986), 3–96, developed a sophisticated systems model to explain technological progress in human history, which Hughes admits to being unaware of until he completed his *Networks of Power* manuscript. See Hughes, *Networks of Power,* 5 (footnote 4). While Hughes's view of system is a complex network of technical and nontechnical entities brought together to solve a problem, Gille's view of system is a complex ensemble of techniques. For Gille, "Prolegomena," 17, all techniques that make up the system have certain internal coherence. The system evolves as the techniques gain certain structural features to attain stability or equilibrium. As Antoine Picon, "Towards a History of Technological Thought," in *Technological Change,* Robert Fox, ed., 37–49, 38, correctly points out, unlike Hughes's notion of a technological system, Gille's conceptualization of a technological system may be difficult to relate to "more socially oriented studies of technology and technological change."

26. Hughes, *Networks of Power,* 7.

27. Hughes, *Networks of Power,* 6–7.

28. Hughes, *Networks of Power,* 7.

29. Although engineers, managers, and financiers are involved in this phase, Hughes, *Networks of Power,* 14, argues that they take up key leadership positions in system building only at later stages. The "mothering" of the infant system rests with the inventor-entrepreneur.

30. Hughes, *Networks of Power,* 14, claims that he borrowed the concept of reverse salients from military historians "who delineate those sections of an advancing line, or front, that have fallen back as 'reverse salients'." Ron Westrum, "The Social Construction of Technological Systems. Review of Bijker, et al. (eds.), *The Social Construction of Technological Systems,*" *Social Studies of Science* 19 (1989): 189–191, contends that Nathan Rosenberg first developed the concept behind "reverse salients" in 1969 in a concept Rosenberg called "technological imbalance." According to Westrum, "Social Construction," 190, reverse salients "may be more exciting, but Rosenberg deserves credit for his earlier very insightful development of what is essentially his idea."

31. Hughes, *Networks of Power,* 14–15.

32. Direct current (D.C.), although a better form of electricity for lighting and traction, was highly uneconomical to transmit long distance due to high transmission loss.

33. The invention and development of high-voltage transformers and high-tension power lines solved these reverse salients, which in turn resulted in an explosive growth of electric power systems.

34. Hughes, *Networks of Power*, 15.

35. The rise of Stone & Webster from a team of lowly installation engineers into a massive electric holding company in the early part of the twentieth century in the United States is an example. Hughes identifies J. P. Morgan as a typical financier tycoon who took advantage of the rise of the electric utility industry, and Samuel Insull as an archetypal utility manager who from being a personal secretary to Thomas Edison became the manager of the General Electric Company in Schenectady, New York, and eventually the system building manager of the giant Chicago Edison Company.

36. Hughes, *Networks of Power*, 17.

37. Hughes, "Technological Systems," 52.

38. Terry S. Reynolds, "Review of *Networks of Power: Electrification in Western Society, 1880–1930*," *Technology and Culture* 25 (1984): 644–647, raises the general concern whether the "system" has been accepted by the historians of technology as a general framework, or is applicable only to Hughes's particular case study. John Law, "The Structure of Sociotechnical Engineering—A Review of the New Sociology of Technology," *Sociological Review* 35 (1987): 404–425, also questions Hughes's proclivity to use the same analytical framework to analyze micro as well as macro phenomena. Because of the ex post facto nature of Hughes's analysis, Law doubts if his model can be applied with equal success to other cases with a different dynamics of system growth and development. My attempt to use Hughes's model to explain the dynamics of the technological change in the Green Revolution, in chapter eight, was only partially successful.

39. Hughes, "Technological Systems," 62.

40. Using the concept of "stasis," Richard F. Hirsh, *Technology and Transformation in American Electric Utility Industry* (Cambridge and New York: Cambridge University Press, 1989), analyzes the state of electric utility industry in the United States since the 1960s when the euphoria over nuclear power began to disappear. Hirsh argues that after having exhausted the benefits of technological advances, system optimality, and economies of scale derived from large-scale systems, the electric utility industry entered into a period of "stasis." Although limited in explanatory power, the concept of "stasis" sheds important light on the process of technological change at the sunset stage of a technological system.

41. Reynolds, "Review of *Networks*."

42. Law, "Sociotechnical Engineering."

43. Edward W. Constant, *The Origins of the Turbojet Revolution* (Baltimore: Johns Hopkins, 1980), "A Model for Technological Change Applied to the Turbojet Revolution," *Technology and Culture* 14, no. 4 (1973): 553–572, and "The Social Locus of Technological Practice: Community, System, or Organization," in *Social Construction*, Wiebe Bijker, et al., eds. (Cambridge, MA: MIT Press,1987), 223–242. Constant admits that two other intellectual influences in his attempt to adapt the Kuhnian model to explain technological change were Karl Popper and Donald Campbell. Constant admits that the evolutionary epistemology of Popper, which was developed fully by Donald T. Campbell, "Evolutionary Epistemology," in *The Philosophy of Karl Popper*, Paul A. Schlipp, ed. (La Salle, IL: Open

Court, 1974), 413–463, was a key inspiration to his model besides Karl Popper's, *The Logic of Scientific Discovery* (London: Routledge, 1992), logic of scientific discovery.

44. Thomas S. Kuhn, *The Structure of Scientific Revolutions* (Chicago: University of Chicago Press, 1970).

45. Constant, "Technological Change," 554.

46. Constant, "Technological Change," 554. This claim, regarding technology, may be defendable. However, given the practice of dismantling the observation-theory, fact-value, empirical-normative, descriptive-prescriptive divide in the construction of knowledge in modern intellectual disciplines, and coming particularly in the wake of the reception of Kuhn's work on imputing radical contingency to scientific knowledge claims, Constant's proclamation of attaining a perfect match between observation and theory in scientific practice may be undefendable.

47. Constant, "Technological Change," 555.

48. Constant, "Technological Change," 555, qualifies this reification of science by interjecting the statement that the "scientific insight upon which the anomaly is founded must be expressible in quantitative form." Quantification of this new revelation is the key factor that can persuade the relevant technological communities to switch to the new paradigm.

49. The turbojet came about as a result of nearly two centuries of turbine development, starting with water turbines, steam turbines, internal combustion gas turbines, and piston-engine turbosuperchargers.

50. Constant, *Turbojet Revolution*, 29.

51. Constant, *Turbojet Revolution*, 31.

52. To be fair, Constant admits this deficiency. He admits that by adopting the paradigm model, he was forced to sacrifice one for the other. That is, his conceptual framework "partially obscures" other critical factors mentioned above. This admission does not rule out the possibility that these factors can be brought in as dependent variables in his model of technological change.

53. Constant, "Social Locus," 224.

54. David W. Lewis, "Review of *The Origin of the Turbojet Revolution* by Constant, E. W.," *Technology and Culture* 23, no. 3 (1982): 512–516, and Edward C. Ezell, "Review of *The Origins of the Turbojet Revolution* by Constant, E. W.," *American Historical Review* 87 (1982): 155.

55. Ironically, Constant rejects the Chicago school of sociologists' (see the next chapter) theory of invention out of hand claiming that it is inadequate to capture the complexity of technological change, and that it is "simply wrong." See Constant, *Turbojet Revolution*, 2. Nevertheless, in the same breath, Constant claims that technological revolutions are initiated by outsiders, which, alas, is similar to principle thirty-two (that inventions are usually carried out by outsiders) of the thirty-eight social principles of invention developed by S. Colum Gilfillan, *The Sociology of Invention* (Cambridge, MA: MIT Press, 1970).

56. George Basalla, *The Evolution of Technology* (Cambridge: Cambridge University Press, 1988).

57. Basalla, *Evolution of Technology*, 30.

58. Basalla, therefore, concludes that models and metaphors used for explaining scientific change, like the metaphors of political revolutions and worldviews that Thomas Kuhn uses to explain the overthrow of one scientific paradigm by its rival paradigm, are inadequate to explain technological change. Thus he would differ with Edward Constant in his usage of Kuhn's paradigm to explain the turbojet revolution in aviation.

59. The Indian *charka* consists of a pair of long wooden cylinders set in a frame, pressed together, and rotated by hand about the longitudinal axis by a crank. The *charka* was a simple but elegant technology that Mahatma Gandhi tried to propagate as a simple and "appropriate" technology that could make Indians self-sufficient. He urged Indians to go back to their own technologies to create their own self-sufficient republic instead of imitating "modern" western technology. The *charka* formed a potent symbol of resistance against British imperialism. Gandhi's symbol of *charka,* however, did not survive in the postindependent India of Nehru and other modernizers.

60. Basalla does not quite explain how this technology reached the Louisiana Territory from the Levant.

61. Basalla, *Evolution of Technology,* 32–34.

62. Basalla, *Evolution of Technology,* 32–34.

63. Vacuum tubes evolved from the incandescent lamps. Although Thomas Edison noticed a flow of current from a filament across a vacuum (a flow of current from the cathode to the anode in his earlier vacuum lamps), he did not understand what caused the phenomenon. This was largely because electrons were not discovered yet. John Ambrose Fleming utilized the directional flow of current from the filament to develop a diode. Later, Lee de Forest added a third element to regulate the current flow from the filament by changing the voltage of this element to produce a triode. De Forest called his device an "Audion" and claimed priority for the discovery of the radio, although he did not understand the electronic principle behind his device. Irving Langmuir of GE and H. D. Arnold of AT&T first provided the electronic interpretation of the "Audion." For more on the development of the radio and the protracted legal battles involving three key individual developers of the radio (Fleming, de Forest, and Edwin Armstrong) and three powerful corporations, GE, AT&T, and RCA, see Hugh G. J. Aitken, *The Continuous Wave: Technology and American Radio, 1900–1932* (Princeton, NJ: Princeton University Press, 1985). The incandescent lamp can be considered the precursor of vacuum tubes.

64. Basalla, *Evolution of Technology,* 43–46.

65. Newton Copp and Andrew Zanella, *Discovery, Innovation, and Risk: Case Studies in Science and Technology* (Cambridge, MA: MIT Press, 1993), 131.

66. Copp and Zanella, *Discovery, Innovation.*

67. Basalla, *Evolution of Technology,* 63.

68. Basalla, *Evolution of Technology,* 218.

69. John K. Smith, "Thinking about Technological Change: Linear and Evolutionary Models," in *Learning and Technological Change,* Ross Thomson, ed. (New York: St. Martin's Press, 1993), 65–78.

70. An excellent example to corroborate this claim was provided by Michel Callon in which he narrates the failure of the electric car (VEL) in France in the 1970s. For more on this case, see chapter three.

71. Smith, "Technological Change," 69.

72. The actor network models of Bruno Latour, Michel Callon, John Law, and others circumvent this problem by attributing agency and reflexivity to all entities (animate or inanimate) associated with a technology. For more details on the actor network models, see chapter three.

73. Price, "Historiographic Revolution."

74. Barry Barnes, "The Science-Technology Relationship: A Model and a Query," *Social Studies of Science* 12 (1982): 166–172, and Walter G. Vincenti, *What Engineers Know and How They Know It* (Baltimore: Johns Hopkins, 1990).

75. Edwin T. Layton, "Through the Looking Glass; or News from Lake Mirror Image," in *In Context: History and History of Technology,* Stephen H. Cutcliffe and Robert C. Post, eds. (Bethlehem, PA: Lehigh University Press, 1989), 29–41, and "Technology as Knowledge," *Technology and Culture* 15, no. 1 (1974): 31–41; and Vincenti, *Engineers Know.*

76. Layton, "Looking Glass," 35.

77. Vincenti, *Engineers Know,* 11.

78. John M. Staudenmaier, *Technology's Storytellers: Reweaving the Human Fabric* (Cambridge, MA: MIT Press, 1985), 103.

79. Vincenti, *Engineers Know.*

80. Vincenti, *Engineers Know,* 17.

81. Constant, *Turbojet Revolution.*

82. Davis's airfoil worked, but not for the reasons he gave. It was a low drag design similar to some existing NACA shapes. Boeing adopted a similar high aspect ratio design for the B-29 after an engineer brought the idea from Consolidated. Davis did not have a "method" of designing airfoil. I am grateful to Robert Ferguson for this information.

83. Hayden White, *Tropics of Discourse: Essays in Cultural Criticism* (Baltimore: Johns Hopkins, 1985).

84. See especially, Max Weber, *The Protestant Ethic and the Spirit of Capitalism,* tr. Talcott Parsons (London: Harper Collins Academic, 1991), and *General Economic History* (New Brunswick, NJ: Transaction Books, 1981).

85. Hård, *Machines are Frozen.*

86. Hård, *Machines are Frozen,* 241.

87. White, *Tropics of Discourse,* 51. According to White, the theoretical frame of a particular emplotment has a corresponding ideological implication. Some modes of emplotment and their corresponding ideological implications are romance-anarchist, comedy-conservative, tragedy-radical, and satire-liberal. White also claims that the historical narratives of metahistorians can be read as stories or literary texts by using appropriate tropes, such as metaphor, metonymy, irony, and synecdoche.

88. In a letter to a young zoologist, John Scott, Darwin advised him: "let theory guide your observations" (quoted in Peter Novick, *That Noble Dream: The "Objectivity Question" and the American Historical Profession* [Cambridge and New York: Cambridge University Press, 1988, 36]).

89. Oswald Spengler, *The Decline of the West,* 2 volumes (New York: Knopf, 1992), 31, emphasis in original.

90. For more on the identification of these as an "axiomatic core," see Louis Girifalco, *Dynamics of Technological Change,* 13.

91. Karl Marx, *Capital, The Poverty of Philosophy* (Moscow: Progress Publishers, 1955), and *Capital.* Karl Marx and Friedrich Engels, *The Manifesto of the Communist Party* (Chicago: Encyclopaedia Britannica, 1952).

92. This is his greatest work on comparative world history, spanning 12 volumes (1934–61) and the thirteenth volume in 1972 (all published by Oxford University Press).

93. C. T. McIntire and Marvin Perry, "Toynbee's Achievements," in *Toynbee: Reappraisals,* C. T. McIntire and Marvin Perry, eds. (Toronto: University of Toronto Press, 1989), 3–31, 10.

94. McIntire and Perry, "Toynbee's Achievements," 10.

3

Sociological Models

FROM SOCIOLOGY OF SCIENCE
TO SOCIOLOGY OF TECHNOLOGY

Despite an early blossoming of theoretical and empirical explorations on techno-logical change, curiously sociological interest on technology in general, and tech-nological change in particular, went into a dormant stage until the 1980s and 1990s. However, despite the temporary hiatus, the comeback was very significant, as recent sociology-inspired studies of technological change are intellectually en-gaging and most productive. Like in other disciplines and areas of inquiry, the so-ciology of technology evolved as an extension of the theories and models of soci-ology of science *to* technology. This was again a reflection of the emergence of the sociology of scientific knowledge (SSK) as the dominant paradigm in the so-ciology of science of the 1970s with the Edinburgh Science Studies Unit acting as its intellectual fountainhead.[1] The Edinburgh program treated science as any other cultural artifact or belief system. For the followers of the Edinburgh program, sci-ence enjoys its high status not because it has a monopoly on truth claims about na-ture and natural processes, but because of its spectacular success in solving many of society's problems through various practical applications.

The "Strong Programme" in the sociology of science spearheaded by David Bloor gave the greatest theoretical support to SSK by claiming that, in matters of assessing truth claims about scientific theories, one must maintain symmetry.[2] That is, one must resort to the same sociological explanation to all belief systems irrespective of their truth claims, or the same method of analysis must be used to explain a true scientific theory from a false theory. The SSK proponents also hold that there is no such thing as a demarcation problem in science where any effort to demarcate nonscience or pseudoscience from science proper would be futile. For the proponents of SSK, in fact, there is no such thing as pseudoscience. This is because, as per SSK canon, science does not provide its own methodology for appraising its credentials as a separate cultural activity that is distinct from the practice of other cultural pursuits. This theoretical breakthrough in the sociology of science found its operational success as a result of the emergence of the relativist thesis called the "Empirical Programme of Relativism or EPOR" propounded by Harry Collins.[3] Closely following these developments in the social studies of science, the social studies of technology evolved as the extension of these theoretical breakthroughs in SSK. In sociology of technology, as in sociology of science, the concept of symmetry should be upheld to explain successful technological innovations as well as failed ones.

Technological development should be interpreted according to the particular contextual situations of the community of its users and developers. The success of a technological innovation must be analyzed in the same analytical framework as a failed innovation. As Trevor Pinch puts it, "in the study of technology we do not want to fall into the trap of assuming that successful technological products and processes require no explanation."[4] But as we can amply verify from the literature on technological development, most models and theories of technological change are, however, based on successful inventions and innovations. Nevertheless, better explanatory power for models of technological change evolve, from testing the models against failed inventions and innovations, as will be shown later.[5] As Pinch, in turn, observes, "[O]nly when the failed development [of technology] is recovered can the interesting, explanatory questions be asked."[6] The social constructivists take it as axiomatic that the success or failure of a technology cannot be simply reduced to a functionalist account of its workings or its internal structures, but its achievements and problems have to be explained in terms of social factors alone.

In this chapter, several different empirically based theories and models of social construction of technology (SCOT) that evolved originally from the social studies of science will be analyzed. The most interesting works in the social construction of technology frame, such as the original SCOT model of Pinch and Bijker, heterogeneous engineering, systems, and actor network models of technological change, fall within the category of what Sergio Sismondo calls "radical" social constructivism.[7] However, we will analyze one "mild" version of social constructivism called the political economy of technological change, to map out all the possible versions of the social construction of technology.[8] Before getting onto this recent scholarship in the social studies of technology, let us review some early sociological attempts to understand invention, innovation, and technological change.

CLASSICAL SOCIOLOGY OF TECHNOLOGY

It is usually reported that Karl Marx was the first social scientist to provide in his social theories a rigorous explanation and a historical synthesis of technological change.[9] However, a little-known contemporary sociologist of Marx, Eilert Sundt, also offered an interesting account of technological change that was mainly inspired by Darwin's work on the evolution of species. Sundt was a Norwegian theologian by profession, who wrote several interesting works on demography and sociology of technology.[10] The two empirical cases on which Sundt built his theory of technological change came from house construction and boat building in Scandinavia. According to Sundt, technological variations, in the beginning, are random occurrences because no boat builder can build two boats exactly the same. The user may eventually notice this "accidental" change (variation) and recommend that any contingent progress should be incorporated in future boats. However, this "variation-cum-selection is a *local* maximum"[11] as further improvements of the boat may result in defects. Thus, what began as an imperfection of the craftsman, his inability to make perfect copies, cumulated in the form of a technological improvement. Sundt's explanation of technological change is the first known application of the evolutionary model to explain this process by looking at the community of technology practitioners as the focusing device.

After these interesting sociological observations on technology in the nineteenth century, no serious scholarship on the sociology of technology emerged until American sociologists William Ogburn[12] and S. Colum Gilfillan[13] offered meaningful interpretations of the dynamic interplay between technological invention and diffusion, and social change during the early to mid-part of this century. Before Robert Solow[14] interpreted Simon Kuznet's extensive data on American economic growth as an epiphenomenon of the technological change that American society had experienced (see chapter four for more details), Ogburn argued that the improvement in the standard of living indicated by Kuznet's data was directly related to the rapid technological change that had taken place in America.[15] Ogburn claimed that three important technological factors contributed to the increase in the standard of living. The first was reflected in increases in the annual number of patents granted (which he used as an important proxy for technological change). A second contribution was the increase in the value of "reproducible equipment in industry and on farms." Finally, Ogburn maintained that increases in the production of energy for improving the mechanization of work and production caused the technological change that resulted in the improved standard of living of the people in the United States.

Attributing Marx as a cohort of the same idea, Ogburn claims that "[T]he inventional interpretation of history is, indeed, like the economic interpretation of history, only one step removed."[16] Although Ogburn uses "invention" to describe his model, he uses it as a portmanteau for all technological activities, from invention to development to diffusion. Invention for him includes all mechanical inventions, inventions in "applied science," and "social inventions." Invention is as much a cultural and social process for Ogburn as it is evolutionary (akin to evolutionary changes in nature). In short, technology is treated as the manifestation of the "material culture" of societies. Technological environment, according to Ogburn, is a "huge mass in rapid motion" with which social institutions find it hard to catch up with.[17] Thus, he says there is always a cultural lag between technological change and social change. Social institutions and cultures always have to adjust to technological change because the institutional mechanism through which material culture is manifested is indirect and roundabout.[18]

The inventive process is one in which the known material and the cultural elements of a society are combined together. Ogburn rejects the heroic invention theory, and instead posits a theory of the cumulation of small changes as the ideal explanation of the "inventional" process. Ogburn rightly points out that many inventions and discoveries are contingent, and that the popular belief that necessity is the mother of invention, a demand theory of invention, is only partly true. He points out that the process of technological change cannot be decoupled from the process of "cultural evolution," and there are four essential factors responsible for this: invention, accumulation, diffusion, and adjustment.[19] Ogburn's theory is counterintuitive and interesting. However, not having provided an empirical case to support it, I believe that one may cast doubt on the use value of his model.

Attempting to develop a "sociology of inventions," in the classical tradition, Gilfillan, another significant sociologist who reflected on technology, widens the frame of reference for the analysis of technological change to nontechnical factors that interact with changes in technical factors. The case study he uses to explain his model of technological change is the "invention" of the ship. Wealth, education, population, and "industrialism" are some of the nontechnical factors Gilfillan adds to the "milieu" of "social principles of invention." Gilfillan presents thirty-eight such social principles of invention. Technological change, according to him, is very much similar to a Darwinian evolutionary process that he subsumes under the "inventional" process. Gilfillan uses the evolutionary theory as more than a heuristic, and instead claims that "[T]he ship or any invention is a biologic [*sic*] organ, in the same sense that a bird's nest is."[20] It is not quite clear how the analogy should be interpreted. While some of the principles of invention are insightful, a few are controversial, and the rest are straight generalizations of ordinary observations from everyday life.[21] After these sociological observations in the nineteenth and early to mid-twentieth centuries, serious reflections on technological change from sociologists declined until the 1970s, when sociology of science and later social studies of technology emerged as significant intellectual components of science and technology studies.

SOCIAL CONSTRUCTION OF TECHNOLOGY

Nearly fifty years had elapsed since Ogburn's and Gilfillan's work before sociologists began to show interest in explaining technological change. The "turn to technology"[22] came as some sociologists of science of mostly European origin looked to the Edinburgh school of the sociology of scientific knowledge (SSK). SSK again was inspired by the "Strong Programme" in the sociology of science of David Bloor[23] and the "Empirical Programme of Relativism" (EPOR) of Harry Collins.[24] For the social constructivists of the SSK tradition, the social construction of scientific practice and knowledge creation forms the model for understanding technology.[25] That is, the foundational principle is the Bloorian axiom of symmetry that requires the analyst to adhere to the same sociological explanation, methodologically speaking, for all beliefs independent of their truth status. Thus, Pinch asserts that the analyst should not fall into the trap that successful technological innovations are self-evident processes and need no explanation, as opposed to failed innovations, which always require a special kind of explanation.[26] Thus the "success or failure of a process or artifact is a social achievement to be explained in terms of social factors."[27]

The social constructivist approach assumes that the construction of artifacts and technological practices are underdetermined by the natural world, and their constructions are socially mediated and determined. Thus, as in science, controversies, friction, and differential power distribution between actors involved during the innovation process are the units of analysis for a social constructivist approach to understanding technological change.[28] Trevor Pinch and Wiebe Bijker see the developmental process of a technological artifact as an "alteration of variation and selection" in a "multidimensional" mode.[29] The central conceptual anchor of their argument is the idea of "interpretive flexibility" of artifacts. This assumption holds that there is more than one interpretation for the "sociological facts" behind the "social construction" of artifacts. Thus, the focus of explanation moves from the strictly internal technological realm of the artifact to the external social milieu. This property of the artifacts is used to explain the process of "closure" through which the artifacts are "stabilized" by the "interests" and "actions" of the relevant social groups.[30]

Pinch and Bijker developed their sociology of technology by treating "technological knowledge in the same symmetrical manner that scientific facts are treated within the sociology of scientific knowledge."[31] The schematic of the particular version of social construction of technology (SCOT) of Pinch and Bijker follows the ensuing arguments. In the case of science, scientific facts are interpreted with respect to nature. In the case of technology, they argue, technological artifacts are interpreted with respect to culture. Pinch and Bijker develop their model based on a detailed historical case study of the development of safety bicycles of the 1890s that evolved from the unstable high-wheeler bicycles of the 1860s. The four theoretical anchors of their model are (1) the *relevant social groups* who exercise, (2) *interpretive flexibility* in the design characteristics of the artifact (bicycles), (3) *closure,* which means the (4) "*stabilization* of the artifact and the disappearance of problems" so that the "final" technological product (safety bicycle) emerges as the culmination of the entire process of technological change. They identify two types of closure: the first is rhetorical closure. In this case convincing the concerned social group(s) that their problem has been solved, which is effected through rhetorical means, such as facile advertisements and other rhetorical tactics, solves a technological controversy. That is, it is not necessary to solve the problem in the common sense way of a technical problem-solving activity. The idea is to convince the concerned social group(s) that there exists no critical problem for the bicycle's acceptance.[32] The second type is closure by redefinition of the problem: closure and acceptance of the technology is achieved by redefining the original problem and finding a solution to another related problem. The success of the closure depends upon the particular sociological and cultural uniqueness of the social groups concerned about the technological problem. The strategy is to identify the key social group(s) first and "enroll" them in the new scheme that would encourage others to follow the enrolled group(s).[33] Pinch and Bijker conclude that the content of an artifact is described by the "meaning" ascribed to it by the social groups, which (meanings) themselves were influenced by the wider society. Bijker in a reformulation of their model presents a nine step "staircase argument," beginning with the prehistory (setting the stage) of the bicycle starting at the bottom of the staircase (step one).[34]

The first viable bicycle that evolved, after a long trial-and-error process, was the high-wheeled "Ordinary," which is identified as step two of the staircase. By tracing the users of the Ordinary, Bijker develops the concept of "relevant social groups" on step three. Focusing on the problems associated with the Ordinary and finding solutions to these problems form step four. The tricycles and the safety Ordinaries that evolved to solve the safety and stability problems of the Ordinary form step five of the staircase. Step six is called "interpretive flexibility" where the conceptual issue is how the artifacts are constituted by the relevant social groups. Step seven of the staircase was characterized by the evolution of the air tyre that emerged as a device for mitigating antivibration and speed enhancement. This stage involved the competition between the safety bicycle and the Ordinary. Step eight is called "closure" and "stabilization," a stage during which the safety bicycle was selected by the relevant social groups over the Ordinary. The final step (nine) where the safety bicycle was accepted by almost all the members of the relevant social groups as a commercially successful technology is identified by Bijker as the "social construction of the safety bicycle" stage.[35]

Attempting to provide a comprehensive "theory of invention," Bijker[36] extends the SCOT program of Pinch and Bijker[37] by introducing two new theoretical concepts, "technological frame" and "inclusion." The case study used for this attempt was the invention of Bakelite as a successful synthetic plastic at the beginning of this century. The concept of technological frame, according to Bijker, "is intended to apply to the *interaction* of various actors. Thus, it is not an individual's characteristic, nor the characteristic of systems or institutions; frames are located *between* actors, not *in* actors or *above* actors."[38] According to Bijker, this theoretical concept should be looked at as the anvil of a successful explanation of the construction of an artifact encompassing the entire social processes from invention to the commercialization of the technological product. Technological frames essentially reveal how existing technology "structures" the social environment. As a result, competing technologies can have multiple technological frames. Technological frames are similar to a Kuhnian paradigm or disciplinary matrix, but they differ in the sense that the community of technology practitioners includes nontechnical people as well, unlike its scientific counterpart where the community is made up entirely of scientific practitioners who are wedded to the same paradigm.[39] The interactionist concept of technological frame, according to Bijker, is what makes it a novel concept, as is also the fact that the same actors can be included in different technological frames. The technological frame is what "structures the interaction among the actors of a relevant social group."[40] Bijker, further, adds, "[A] technological frame comprises all elements that influence the interactions within relevant social groups and lead to the attribution of meanings to technical artifacts—and thus to constituting technology."[41] What makes one frame to become successful over its rival is the process of inclusion or enrollment of more powerful actors. Thus, the selection of Bakelite over Celluloid as the successful synthetic plastic came about because of the enrollment of two powerful actors, the radio and the automobile industries, in the "actor network," about which more to follow later.[42]

The concept of technological frame is a successful theoretical device to analyze and explain the course of technological change within the SCOT framework. Bijker provides the following "tentative list" of elements of a technological frame: goals; key problems; problem-solving strategies; requirements to be met by problem solutions; current theories; tacit knowledge; testing procedures; design methods and criteria; user's practice; perceived substitution function; and exemplary artifacts.[43] As Bijker concedes, this list may tend to make a technological frame a "catchall concept." However, his explanation of the "construction" of both Celluloid and Bakelite include the material, social, and cognitive elements involved in the network of these two "exemplary" artifacts. Similar to the nine step staircase argument developed for the safety bicycle, Bijker develops a staircase argument for the concept of technological frame. He did this by a reverse engineering (adaptation) process by deconstructing the historical narrative of synthetic plastics with Bakelite as the culmination of a successful outcome of a search process.[44]

Besides Pinch and Bijker, there are several others who use the general SCOT framework to analyze technological change. Using an empirical example from the telecommunications industry, the modernization of telephone exchanges in Britain, John Clark, et al.,[45] employ the social constructivist approach to explain the technological change in this emerging high technology area. The modernization of telephone exchanges in Britain started in the 1960s when the existing electromechanical exchanges began to be replaced by the semielectronic TXE range of exchanges.[46] The case involves British Telecom's major decision to invest heavily in this technology, and how managers, trade unions, and work groups shape the process and outcome of this technological change. The decision to introduce the new technology was based on management's "strategic" decision that semielectronic exchanges would be more efficient and cost effective in terms of maintenance costs. Once the new innovation was introduced, the trade unionists and workers tried to "negotiate" with management on ways to incorporate the change in the work environment with minimum disruption. Thus, the process of technological change occurred in stages, such as initiation, decision to adopt, system selection, implementation, and routine operation. During the process of change, substantive "issues" emerged, requiring decisions by the "actors" (managers, trade unions, and work groups) to find "closure" to the problems resulting from these issues. The social construction of the technological change results as a way of addressing the "critical junctures" when the temporal stages and the critical issues interact, thus allowing the actors room for "interpreting" issues related to the technology on a "local" basis.[47]

SYSTEMS AND SOCIAL CONSTRUCTION OF TECHNOLOGY

Staying within the SCOT framework, Donald MacKenzie[48] and John Law[49] explain technological change using the systems approach of Thomas Hughes (see chapter two). Using nuclear ballistic-missile technology as his empirical focus, MacKenzie argues that technological development cannot be understood without a clear knowledge of the underlying organizational, political, and economic issues that ultimately shape the development of technological knowledge related to missile technology. According to MacKenzie, the technological process of attaining extreme accuracy of ballistic missiles is entirely social. Regarding the social construction of missile accuracy, MacKenzie argues that the best framework that will incorporate the three complex issues, categorized as organizational, political, and economic, is the technological systems approach, the approach followed by Thomas Hughes for explaining the evolution of large technological systems. Consequently, for MacKenzie, missile accuracy is attained when the actors become increasingly capable of detecting the "reverse salients" to solving the "critical problems," which is similar to the method Hughes adopted in the electric power systems where he traced the growth of the electric light and power utilities.[50] In missile technology, the guidance system is the crucial part, and as a result, accuracy comes from designing the most advanced inertial guidance and navigation system.[51] The "reverse salient" occurs when the inertial guidance system lags behind as accuracy is reconfigured from kilometers to a few meters of the target. MacKenzie traces the development of ballistic missile accuracy by relating to the three core components of his social category, viz., organizational, economic, and political. MacKenzie argues that attaining extreme accuracy of a missile may appear to be a purely instrumental issue, but when one traces the technological change in this field, these social factors come to the fore as determinants, and therefore it is legitimate to follow the SCOT framework.

In a similar fashion as MacKenzie, a broadly social constructivist explanation of technological innovation was attempted by John Law[52] using the case of the *British* TRS2 tactical strike and reconnaissance aircraft as his empirical focus.[53] Law combines both the network and sociotechnological systems to construct his theory of technological change in which a technological artifact comes into being through different processes.[54] The first issue to be considered in building a technological system is to know how goals or concerns of the relevant technological community are constituted. This "involves an analysis of the way in which problems are discursively brought into being and successfully imposed on system builders by other actors."[55] The second issue that has to be tackled is the search for "scenarios." The idea is to visualize the ideal world in which a heterogeneous array of actors is to be enrolled to solve the problem "hypothetically."[56] The second step is the mobilization of the resources, and the third step is to "juxtapose" the actors to "fit them together into an array that will form a working system."[57] Finally, the most crucial issue is to analyze the "obduracy" of the elements or actors in question in completing the task.[58] In the case of the TSR2 aircraft, the "relative obduracy" of the elements of the sociotechnological system put together by the network builders seems to have quashed this technical innovation.[59] The concept of symmetry, an important conceptual scheme in the SCOT program, in explaining the failure of a major technological device comes to the forefront in Law's analysis of technological change.

The synthesis of the systems and social constructivist models to explain technological change solves many serious grey areas in the systems model. For example, in the systems explanation, it is not quite clear how to specify the boundaries. The social constructivist approach leaves no ambiguity as the change is reflected through the agency of the interested social groups, being fully cognizant of the technological limitations or possibilities of the particular cases in point as appropriate referents. The network-heterogeneous engineering model analyzed in the next section goes a step further by extending agency to all elements of the network whether animate or inanimate.

ACTOR-NETWORK AND HETEROGENEOUS ENGINEERING

Using the galley, which is primarily a war vessel, as empirical focus, John Law[60] provides a full-fledged actor-network explanation of technological change by staying within the SCOT frame. In this example, Law looks at the *process* which led to the domination of the ocean trade route between Africa and the Middle East and later the monopoly over the spice trade with the Malabar Coast of southwestern India by the Portuguese sailors during the late fifteenth and early sixteenth centuries. Based on the concept of "heterogeneous engineering," Law uses the process of the social construction of the galley to show how the Portuguese, who apparently began their exploration to trade in spices, eventually ended up in dominating the Indian Ocean and the colonizing of many Asian and African countries.[61] Compared to the contemporary ocean-going vessels of the late fifteenth century, the galley was a technologically advanced "emergent phenomenon" that has "attributes possessed by none of its individual components."[62] The galley was a strategic association of wood, magnetic compass, men, wind power, pitch and sailcloth, and cannons and guns. The narrow, sleek and long body of the galley revolutionized sailing by considerably cutting down water resistance. The mounting of a magnetic compass on board allowed the ships to abandon the existing practice of hugging the coastline as a directional guide. Also, a thorough knowledge of water currents made seasonal travelling easier. The use of astrolabes and voltas were a great boon for astronomical navigation, allowing the navigators to return successfully to their native land. This was unlike most previous expeditions that ended up in disaster. Finally, it was the presence of the cannon on board that gave the Portuguese total command of the ocean routes that led to the foundation of a four-hundred-year imperial domination of several Asian, African, and South American waters, territories, and nations.

Law argues that because of the "heterogeneity" of the technological activity, a purely sociological or technological (systems) approach might come close to capitulating to reductionism in explaining the technological change. The essential theoretical point is to understand how stability can be maintained in the construction of the heterogeneous network when hostile and contingent forces threaten the stability of the system. It follows that "the structure of the networks (or systems) in question reflects not only a concern to achieve a workable solution but also the relationship between the forces that they can muster and those deployed by their opponents."[63] In Law's network approach, neither nature nor society has a direct effect unless they "impinge" on the network builder. Therefore, adhering to the symmetry principle of the "Strong Programme," Law argues that the same type of analysis should be attributed to all components, humans and nonhumans. The second cardinal rule of network analysis is that actors are those entities that have detectable reciprocal influence on each other.[64]

Like John Law, Michel Callon also does not invoke a division between nature and society in his actor-network model of the social construction of technology.[65] Arguing that the study of the process of technological change itself is a process of "society in the making," Callon argues that the very claim that one can clearly demarcate social aspects of technological innovation from, say, scientific or technical aspect of it is specious. He argues that social, economic, or political considerations are not added to the scientific and technical factors in the process of technological change as an afterthought. All these factors are present, right from the beginning. The most important theoretical anvil of Callon's model is that the *relationship* between the heterogeneous actors is what makes an actor network, which in turn cannot be whittled down to one single actor or to a network. According to Callon:

> The actor-network can thus be distinguished from the traditional actors of sociology, a category generally excluding any nonhuman component and whose internal structure is rarely assimilated to that of a network. But the actor-network should not, on the other hand, be confused with a network linking in some predictable fashion elements that are perfectly well-defined and stable, for the entities it is composed of, whether natural or social, could at any moment redefine their identity and mutual relationships in some new way and bring new elements into the network. An actor-network is simultaneously an actor whose activity is networking heterogeneous elements and a network that is able to redefine and transform what it is made of.[66]

With the help of a detailed study of the innovation and failed deployment of the electric car (VEL) in France during the early 1970s, Callon presents an extremely interesting case of a nonlinear process of technological innovation. He argues that the VEL project failed primarily because of the wrong model of society its proponents, the engineer-sociologists of EDF (Electricite de France), adopted. Through the analysis of this failed technological development project, Callon shows the importance of adopting symmetry in explaining technological change. The causes for the failure of VEL are as much technical as they are social. Callon's case study of this failed innovation explains more about the dynamics of technological change than the explanations provided so far for successful innovations.

The VEL was conceived by the engineers of EDF five years after the "great cultural revolution" of May 1968 and one year before the international oil supply crisis, which was caused by the Arab oil embargo on nations that supported Israel during the 1973 Arab-Israeli (Yom Kippur) war. The engineers of EDF, being good engineer-sociologists, noticed a ground swell of opinion against conventional fossil fuel-based cars, the symbol of modernity and its attendant industrial civilization. The VEL was to signal not only a change in social structure, but also to offer a solution to the pollution problems caused by the burning of fossil fuels. Accordingly, the engineers decided to enroll an array of heterogeneous actors, both animate and inanimate, to construct an "actor network" in order to achieve this plan. The cast of actors included lead accumulator batteries, electrodes, electrons, electric traction systems, EDF engineers, municipalities (interested in improving public transit), government ministries (concerned about pollution, transportation, and the quality of life of French citizens), Renault car company (potential manufacturer of the chassis of the VEL), social movements, and the environmentally and socially conscious consumers, among other "actants." Despite being successful in enrolling these actors, and despite their knowledge of electrochemistry and politics, the VEL did not take off.

Callon argues that the failure of VEL was not only due to the lack of cooperation from electrons and catalysts, but also due to the wrong model of society that the network builders (EDF "engineer-sociologists") took for granted. The model of society that the engineers used was similar to the social stratification model of Alain Touraine. Touraine's concept of social change is based on class conflict, but unlike Marxists, he argues that we have moved into a "postindustrial" society where the conflict is between the technocrats and their corporate sponsors on one side and the consumers on the other. Consumption is the central conceptual category on which this antagonism unfolds. Although the corporate elite attempts to shape the consumption habits of consumers using false propaganda, and often with facile commercial advertisements, they are constantly on the lookout to produce products that consumers may show a "revealed" preference. But the VEL was not a product that the consumers have shown a revealed preference, and as a result, its failure to capture the market shows that Touraine's model of social change is not adequate.

Callon claims that the failure of the VEL can be explained adequately by Pierre Bourdieu's model of society and social change in the "postindustrial" era. Bourdieu believes that social change is not based on a perpetual confrontation between the ruling (bourgeoisie) and the ruled (working) classes to control technological and industrial power. According to Bourdieu, society is fragmented between various specialist groups, represented by different professional interest groups (politicians, economists, teachers, consumers, and so on), that "maintain mutual relationships of exchange and subordination."[67] However, the society is bound together by a group logic that is sustained by imitating the consumption and behavioral pattern of the upper or elite classes. Despite their inability to hold on to state power directly in liberal democratic societies, the elite still retain enormous power in shaping the behavior and actions of the lower classes. Although they might reject the fact, lower classes constantly vie with each other to imitate the consumption patterns of the rich and famous.

VEL was not targeted at the upper classes as a car of distinction. Driven by a different social logic for finding distinction through consumption, the upper class still chose the expensive traditional cars instead of the utilitarian electric cars. The car embodies a very unique symbol of social standing, as it is more than a mere technological artifact for transportation. Consumers select more differentiation in cars according to their tastes and imitative preferences. Trying to introduce a simple, crude electric vehicle that was still unreliable, meant too optimistic a belief in radical social change effected through technical artifacts. Had the EDF engineers designed VEL as an attractive consumption option for the elite, the lower classes, to whom the VEL was initially targeted, would have bought it had there been a market demand for it among the upper classes. Accordingly, Callon concludes that engineers and technologists who envision transforming society should be aware not only of the latest advances in scientific and technological issues but also of numerous other issues as well. They should also have the correct sociological analysis and concept of social change for the innovation to be successful, in addition to "ironing out the wrinkles" in lead acid batteries and other material devices enrolled in the network.

FOLLOWING THE ACTORS OF THE ACTOR-NETWORK

Following the actors, the relevant social groups, as the *locus classicus* of understanding the process of the construction of facts and artifacts, emerged as the most fruitful ethnographic method for the anthropology of science and technology in the social studies of technology.[68] Bruno Latour's position is that what matters in technoscience is "power," in its multifarious forms. Also, he avers, society and technology are abstract artifacts that are two sides of the same entity. When one is interested in developing the instruments of power, it is crucial to make the connection between the various entities that are emanations of these two abstractions. The result is a network comprised of electrons, microbes, men, women, societal institutions, and so on to complete the mission. The crucial point is to ascertain which constitutional form of the association or network is stronger or weaker than the other alternatives.[69] Latour correctly points out the crucial factor that technological innovation is an unstable event. It is a process constantly in the formation, and measuring it with already defined concepts of efficiency and productivity is only a crude way to accommodate the uncertainty. What is required is to recreate and redefine jaded economic and social vocabularies in order to construct a new network such that it would become a network made up of a seamless array of social and technical elements.

In order for an innovator to develop a new technology, she needs not only to enroll human and nonhuman entities in her network, but also to actively form alliances with all of them, as part of a heterogeneous network.[70] Despite being the inventor of the diesel engine, Latour argues, Rudolf Diesel the inventor failed to transform his diesel engine patent into a workable product because of his inability to enroll the right entities in his network.[71] In his attempt to transform his ideas of a perfect Carnot engine into a workable prototype, Diesel enlisted the help of several engineers and machine tool firms like Maschinenfabrik Augsburg-Nurnberg (MAN) and Krupps. It took several years for Diesel to build a prototype of his "rational" engine. When fuel combustion became a serious problem, he experimented with injecting compressed air that required huge pumps and more powerful cylinders. Finally, the machine became larger, heavier, and too expensive to be able to compete in the marketplace. To solve the marketing problem, Diesel sent blueprints, prototypes, and engineers to anyone interested in paying royalties to his firm on the condition that if his prototype did not work he would return the royalties to the licensees. Because of their inability to replicate Diesel's "black box prototype" into workable engines, all licensees returned his model and Diesel went bankrupt and suffered a nervous breakdown. However, around 1908, a year after Diesel's patent lapsed and became public property, MAN offered a successful engine. Although Diesel claimed later, years before his suicide in 1913, that the new real diesel engine was actually a replication of his earlier model, in reality, the new engine was the outcome of hundreds of engineers' assiduous effort to transform Diesel's dream of a high-compression internal combustion (IC) engine. As a result, despite his name being attached to the engine, we can say that Diesel was merely an element of the actor network of the technological change in the internal combustion engine and modern transportation technology. The story of the diesel engine can be told only by following all the actors, as Latour eloquently shows, with the invention of the diesel engine by Diesel acting as merely a historical footnote.[72]

The operationalization of the Latourian concept of following the actors by remaining within the SCOT framework to explain technological change was further attempted by Wiebe Bijker.[73] According to Bijker, "technological artifacts do not exist without the social interactions within and among social groups."[74] Bijker uses his theoretical concept of "technological frame" (discussed earlier) and "power" to analyze the interactions within and between "relevant social groups," and also to explain the innovation of fluorescent lighting, his empirical focus. For his analytical work, Bijker chose the following four relevant social groups (actors), namely, the social group of Mazda[75] companies, the electric power corporations (electric utilities), the U.S. government (specifically, the anti-trust division within the U.S. Attorney General's office), and the independents (firms that independently developed fluorescent lamps, most importantly, the firm of Hygrade Sylvania). By analyzing the conflicts between them and the process of attaining closure of these conflicts, Bijker delineates how the high-intensity daylight fluorescent lamp became a stable technological innovation that revolutionized artificial lighting.

Although the possibility of fluorescent lighting was found out as a technical possibility decades earlier, it was only in 1938 that the Mazda companies commercially released a low-voltage fluorescent lamp.[76] The first commercially successful fluorescent lamps used high voltages to generate almost any color of the spectrum by bombarding the ultraviolet light produced by low-pressure mercury vapor on fluorescent powder inside the tube.[77] However, factors such as the color of the light, the high voltage, and the expensive fixtures required limited the application of this fluorescent lamp to outdoor use only. Although high-voltage fluorescent lighting was available for indoor lighting in Europe, a European company's (Claude) effort to diffuse that technology for indoor lighting in America in the late 1930s was blocked by General Electric, which had the monopoly on incandescent lamp business in the United States. Through a licensing arrangement, General Electric allowed Claude to market outdoor, high-voltage fluorescent lighting while General Electric kept a monopoly on indoor high-voltage fluorescent lighting. Noticing the imminent possibility of a low-voltage fluorescent lamp dominating the artificial lighting market by independents (notably Hygrade Sylvania Corporation), the Mazda companies (basically General Electric) developed and began to market low-voltage fluorescent lamps to the extreme displeasure of electric utilities (the reasons for this displeasure anon). However, despite this, the Mazda group in cahoots with its hitherto partners in business, the utility companies, shut out the independents from entering the artificial lighting business.

The conflict between the Mazda companies and electric utilities revolved around the low power factor of the low-voltage fluorescent lamps. The utilities complained that the low power factor of the low-voltage fluorescent lamps would overload the distribution system, and also would create revenue loss because of the reactive power of this lamp (the cost of which cannot be charged to the consumer). The utilities prevailed upon the Mazda companies to abandon the project. In addition to the low power factor, the utilities were also worried about the lower power demands by household consumers if they switched over to low-voltage fluorescent lamps from incandescent lamps because the fluorescent lamps would consume less energy than their incandescent counterparts. However, the introduction of high-efficiency daylight fluorescent lamps by an outsider, Hygrade Sylvania, challenged the two dominant players, the Mazda companies and the utilities, from monopolizing the market. The Mazda companies saw a threat to their dominance over the lamp market, while the utilities were concerned about their load factor management. The high-efficiency fluorescent lamps would consume less power, hence could lower power demands that could complicate load factor management, and eventually would threaten the monopoly profits of this load-dominant social group. A partial closure to this controversy, which threatened the status of the two dominant players, was effected by enrolling another relevant social group, the fixture manufacturers. Their enrollment could solve two problems for both players: (1) shut out the independents, specifically Hygrade Sylvania, from the fluorescent lamp market by the utilities demanding certification for auxiliaries to exclude this new player from dominating the lamp market; and (2) the utilities could correct the power factor problem by having the fixture manufacturers install power factor correction technology incorporated into the fixtures. The closure to the controversy came as the Mazda companies introduced the high-intensity daylight fluorescent lamps in 1942, which shut out the outsider, at least temporarily.[78]

This is a very interesting case of a technological change where the consumers, the government, and other significant social groups play only a peripheral role in shaping technological change. The United States government's efforts to break up the monopoly on the artificial lighting business by the cartel formed between the Mazda companies and the utilities was foiled by the immense power of General Electric, which argued that government interference would affect the United States' war preparedness as the company was involved in the production of many war products during the Second World War. They argued that any antitrust case should be postponed until after the war. This waiver, however, consolidated their power, effectively shutting out the outsiders. The invention of high-efficiency daylight fluorescent lamps by Hygrade Sylvania should have benefited the public in terms of low energy cost. But the utility companies went on a high profile advertising and propaganda blitz to argue that Sylvania's claim about high efficiency was not true and tried their best to keep Sylvania's high-efficiency fluorescent lamp from becoming the most viable artificial lighting technology of the time. The Mazda companies and the utilities curbed the popularity of the Sylvania high-efficiency fluorescent lamps by introducing certification for fluorescent lamp fixtures, and later by developing their own high-intensity fluorescent lamps.[79]

The social construction of the development of high-intensity fluorescent lamp is a clear example of how powerful social groups try to influence the course of technological change to satisfy their short term business interest at the expense of long term public interest. Bijker convincingly shows the role of "power" in the construction of the fluorescent lamp technology in the United States by arguing that there is a clear relationship between the "social shaping of technology and the technical shaping of society."[80] Bijker shows how power can determine the shape of artifacts, or to be precise, what artifact becomes successful by linking social construction with "an interactionist conception of power." Bijker divides the interactionist concept of power into two subconcepts: (1) the semiotic nature of power, and (2) the micropolitical

nature of power. The semiotic nature of power is fixed and represented in technological frames, while the micropolitical nature of power is directly linked to the closure and stabilization processes.[81] Apparently, these two power processes do not act
independently, but in a dialectical manner, as Bijker explains:

> The reaching of closure, whereby the interpretive flexibility of an artifact is dimin-
> ished and its meaning fixed, can now be interpreted as a first step in constituting semi-
> otic power, resulting from multitude of micropolitics to fix meanings. In the subse-
> quent stabilization process further interactions result in fixing more elements into the
> semiotic structure—enlisting more people in the relevant social groups, elaborating
> the meaning of the artifact. A technological frame then constrains actions of its mem-
> bers and thus exerts power through the fixity of meanings of, among other elements,
> artifacts; this is the semiotic aspect of the new power conception. A technological
> frame also enables its members by providing problem-solving strategies, theories, and
> testing practices, for example, which forms the micropolitical aspects of power.[82]

Although the artifact in his model circumscribes the distribution of power, Bijker
gives agency to the artifact. In this manner the agency that the artifact gained is
actually the agency vested in the social power of the dominant social groups. Although Bijker concedes that his conception of power should not be interpreted as
a novel theoretical concept, what he asks is to read it along at the explanatory level
using the theoretical premise of the SCOT program.

SOCIAL CONSTRUCTION AND POLITICAL ECONOMY

Despite the high success enjoyed by the empirical works in the social studies of
technology discussed earlier in explaining technological change, one of the serious concerns raised by its detractors like Langdon Winner[83] was the question of
relevance to society in terms of its usefulness in bringing about meaningful social
change. The detractors argued that the social studies of technology did not provide
a radically new interpretation of technological change to effect meaningful social
change because all it did was to provide a purely determinist account of technological change with either technology or society as the independent variable.[84]
One of the major deficiencies noted in the social constructivist approaches was the
relegation of the political economic aspects of technological change to a peripheral level.[85] This concern becomes particularly salient when analyzing the impact
of modern technology on the Third World and the dynamics of technology transfer from the industrialized nations to less-developed nations. The question as to
whether the tenet of symmetry can be applied to two historically divergent and socially, culturally, and economically disparate systems in a postcolonial context is
a serious issue that social constructivist models need to address.

Apparently being sensitive to the above concerns, Steven Yearly attempted an
innovative and potentially useful approach to studying technology and social

change within the social constructivist paradigm of the SSK tradition.[86] By bringing together the political economy of technological knowledge production, Yearley attempts to synthesize the former with SCOT. The fundamental basis of the
political economic view is that the "development of scientific and technical
knowledge is recurrently shaped by commercial and political priorities" of the predominant forces in society.[87] Although the political, economic, and social constructivist views of the historical development of scientific and technological
knowledge production are opposed to the idealist interpretation of its development, there is an internal contradiction between the ways these two frameworks
view and interpret idealism. Most political economy models of economic development and social change may reflect the principles of idealist thinking by adhering to a specific ideology, whether it is orthodox Marxism or laissez-faire free market economics, although they would overtly criticize idealism as antithetical to
their respective belief systems. However, as Yearley correctly points out, the social constructivist model is patently anti-idealistic because of its obvious adherence to relativism. Yearley tries to bridge the gap between the social constructivist
and political economic views by concentrating on their mutual antagonism to the
idealist interpretation of knowledge production and social change.

The analysis of the political economic view of technological change and economic development was, based on the historical experience of the modernization
project, embarked on by the Third World after decolonization. Yearley strongly
opposes the promotion of the western "natural" model of modernization theories
for the Third World as historically unacceptable and untenable. He argues that the
industrialized countries "developed" because they had the opportunity to exploit
the developing Third World countries, and the Third World is not favorably situ-
ated to repeat the historical experience of their colonizers. Yearley rejects the functionally equivalent modernization theories of Weber and modern-day neoclassical
economic development theories of such figures as Walt Rostow[88] and Arthur
Lewis.[89] Instead, Yearley lays the groundwork for his model based on dependency
theory to explain how modern scientific and technological knowledge have become inadequate, and often counterproductive, to bring about economic development to the Third World.[90]

Dependency theory may be described briefly as follows. The decolonized nations' (collectively put as the "periphery") economic and political structures are
shaped by the involuntary economic and political relationship they had with their
erstwhile colonizers ("core" industrialized nations). "Specifically," as Yearley
puts it, "this means that their [periphery's] economic institutions and infrastructure have been structured in response to the demands of the core nations."[91] The
artificially created dependence of the periphery on the core creates underdevelopment in the former. The cause of underdevelopment, according to dependency theorists, is directly linked to the dependent relationship. Also, the underdevelopment
of the periphery is exacerbated by the development of the core, as the former is a
necessary condition for the latter. Yearley concludes that the technology transfer

and aid programs create a dependent science and technology in the Third World. Third World technological knowledge production sites are extensions of the First World centers in order to serve the interests of the latter. Yearley claims that the Green Revolution in Third World agriculture is a classic example of a technology that evolved out of the dependent relationship between the core and the periphery. The failure of this technological fix to alleviate poverty and deprivation is analyzed within the contexts of social constructivism and political economy. The claim that the Green Revolution in India is the result of a dependent relationship may be only partly true as it shows a much more complex picture of a nonlinear case of technology development than a linear case of dependent technology development. A detailed case study of the Green Revolution and analysis of it using different theoretical angles are attempted in chapters six, seven, eight, and nine.

CONCLUSION

The broadly conceived social constructivist approaches discussed in this chapter follow the empirical program of "social shaping of technology" as the underlying theoretical premise of explaining technological change. Despite the caveat that one must not get into the determinist trap either way, technology does not determine social change nor does society determine technological change unilaterally. But that is a contradiction from the point of view of a strict social shaping of technology framework. What remains to be argued in such a theoretical framework is that everything is social through and through. In essence, to be technological is to be social.

According to Bijker and Law, the common thread of social studies of technology is "heterogeneity and contingency."[92] Thus, technological change unfolding according to a priori internal logic is abandoned, because it is impossible to reduce this phenomenon into a formalizable theoretical explanation. Technological change is not only contingent, but an emergent phenomenon as well. It arises out of conflict, confrontation, and planning. The social constructivists attribute extreme interpretive flexibility to technological artifacts, and thus, technology can be stabilized only when the "heterogeneous relations in which it is implicated, and of which it forms a part, are themselves stabilized."[93]

Social constructivist explanations of technological change vary from the original SCOT model of Pinch and Bijker, the systems model that was originally developed in the history of technology, the actor network and heterogeneous engineering models, and the political economic model. These models can be broadly categorized under the "mild" and "radical" versions of social construction of technology, following Sismondo's interesting discussion of different meanings of social construction in the sociology of science and technology.[94] Unlike the interest, systems and political economic models, the network theorists do not believe in a social, economic, or cultural backdrop with which the elements of the network are contrasted or compared. The backdrop itself forms part of the network.[95] Thus the network builders attribute agency to all elements of the network, whether they are humans or electrons. Structures and materials should be treated as individual social entities, and these in turn should be treated symmetrically as human entities in order to tell the complete story of technology in/and society.[96] That is, inanimate actors have as interesting a social life as animate (human) actors do.

The actor-network-heterogeneous-engineering model, despite the above problems, comes cleanest in terms of taking a middle ground between the technological determinist explanation of technological change (internal factors or certain inherent logic of technological development determining technological change) and the social determinist explanation (invention, innovation, and other technological activities are based on social needs) of technological change. Against these linear models, the nonlinear SCOT and actor-network-heterogeneous-engineering models look for both social categories (actors) as well as nonsocial categories (actors) to explain technological change. Following the technologists (defined in its broadest sense of participation in the technological development process) and all the heterogeneous elements of the actor network "in action" becomes the primary heuristic for modelling technological change.

ENDNOTES

1. For further elaboration of this point, see Trevor Pinch, "Understanding Technology: Some Possible Implications of Work in the Sociology of Science," in *Technology and Social Processes,* Brian Elliott, ed. (Edinburgh: University of Edinburgh Press, 1988), 70–83.

2. See David Bloor, *Knowledge and Social Imagery* (London: Routledge and Keegan Paul, 1976), for elaboration of the "Strong Programme."

3. Harry M. Collins, "Stages in the Empirical Programme of Relativism," *Social Studies of Science* 11, no. 1 (1981): 3–10.

4. Pinch, "Understanding Technology," 75.

5. Interestingly, this proposition of the SSK practitioners seems to reflect the philosophy of Francis Bacon, whom Ian Hacking, *Representing and Intervening: Introductory Topics in the Philosophy of Science* (Cambridge: Cambridge University Press 1983), 246, calls the "first philosopher of experimental science." According to Bacon, "No one should be disheartened or confounded if the experiments which he tries do not answer his expectation. For although a successful experiment be more agreeable, yet an unsuccessful one is oftentimes more instructive" (Bacon quoted in Hacking, *Representing,* 247).

6. Pinch, "Understanding Technology," 75.

7. Sergio Sismondo, "Some Social Constructions," *Social Studies of Science* 23 (1993): 515–553.

8. Sismondo, "Social Constructions," provides an interesting discussion of various meanings of social construction of science and technology that are found in the sociology of science and technology. However, Sismondo categorizes them under two broad versions, "radical" and "mild" social constructivism. The "mild" version means that the technology we have are being shaped or influenced by such social forces as politics, consumer preferences,

marketing techniques, and so on. The "radical" version of social constructivism, on the other hand, means that the very content of technology, including its shape, design, and functional parameters, is shaped and influenced by social processes and forces alone.

9. Marx's insights on technological change analyzed in the previous chapter may be viewed as sociological as much as historical in nature.

10. My explanation of Sundt's theory of technological change is taken from Sundt's fellow Norwegian Jon Elster, *Explaining Technical Change: A Case Study in the Philosophy of Science* (Cambridge: Cambridge University Press, and Oslo: Universitetsforlaget, 1983).

11. Elster, *Technical Change,* 137, emphasis in original.

12. William F. Ogburn, *Social Change with Respect to Cultural and Original Nature* (New York: Dell Publishing Co., 1966), *On Culture and Social Change: Selected Papers,* O. D. Duncan, ed. (Chicago: University of Chicago Press, 1964), and "Technology and the Standard of Living," *American Journal of Sociology* 60, no. 4 (1955): 380–386.

13. S. Colum Gilfillan, *The Sociology of Invention* (Cambridge, MA: MIT Press, 1970).

14. Robert Solow, "Technical Change and the Aggregate Production Function," *Review of Economics and Statistics* 39 (1957): 312–320.

15. Ogburn, "Standard of Living."

16. Ogburn, *On Culture,* 64. For example, Ogburn interprets the social history of the American Great Plains as greatly influenced and shaped by the invention of three key technologies:the six-shooter pistol, barbed wire, and the windmill.

17. Ogburn, *On Culture,* 85.

18. Ogburn's notion of "cultural lag" is the basis of his theory of social change or social evolution. Although cultural lag explains the causal linkage with many social problems, cultural lag does not by itself explain how society and civilizational forces evolve.

19. Ogburn, *On Culture,* 23.

20. Gilfillan, *Sociology of Invention,* 14.

21. Irene Taviss, "Review of *Supplement to the Sociology of Invention* by Gilfillan, S. G.," *Technology and Culture* 15, no.1 (1974): 136–138, underplays the importance of Gilfillan's model, and instead argues that he is "concerned about demystifying 'the inventor' and about denying any facile technological determinism" in the inventive process.

22. Steve Woolgar, "The Turn to Technology in Social Studies of Science," *Science, Technology & Human Values* 16 (1991): 20–50.

23. Bloor, *Social Imagery.*

24. Collins, "Empirical Programme."

25. Pinch, "Understanding Technology."

26. Pinch, "Understanding Technology," 75.

27. Pinch, "Understanding Technology," 76.

28. The concept of "power" as a crucial factor in the construction of artifacts was added to the SCOT theoretical frame by Wiebe Bijker, *Of Bicycles, Bakelites, and Bulbs: Toward a Theory of Sociotechnical Change.* (Cambridge, MA: MIT Press, 1995), more than a decade after the original SCOT model was developed by Pinch and Bijker in a paper first published in *Social Studies of Science* in 1984. The same paper was reprinted as Trevor Pinch and Wiebe E. Bijker, "The Social Construction of Facts and Artifacts: Or How the Sociology of Science and the Sociology of Technology Might Benefit Each Other," in *The Social Construction of Technological Systems: New Directions in the Sociology and History of Technology,* Wiebe E. Bijker, Thomas P. Hughes, and Trevor Pinch, eds. (Cam-

bridge, MA: MIT Press, 1987), 17–50. The concept of "power" in the SCOT framework is discussed further towards the end of this chapter.

29. Pinch and Bijker, "Social Construction," argue that their model operates in a multidimensional mode to remedy the "linear" models used explicitly in innovation studies and implicitly in much of history of technology. They also claim that the evolutionary epistemology assumption they use is compatible with the evolutionary epistemology of Stephen Toulmin, *Human Understanding,* Volume I (Princeton, NJ: Princeton University Press, 1971), and Donald Campbell, "Evolutionary Epistemology," in *The Philosophy of Karl Popper,* Paul Arthur Schlipp, ed. (La Salle, IL: Open Court, 1974), 413–463.

30. "Closure" occurs when social mechanisms limit the interpretive flexibility and allow for technological controversies to be terminated before they are resolved.

31. Pinch and Bijker, "Social Construction," 24.

32. In the case of the early high-wheeler bicycles, the safety concerns of the users were allayed not by lowering the wheels or the seating arrangement, but by "facile" advertisements claiming that the high-wheelers are "almost absolutely safe." See Pinch and Bijker, "Social Construction," 44.

33. In the bicycle case study, Bijker and Pinch use the innovation of the air tyre on bicycles first introduced by Dunlop. In the beginning, every concerned group guffawed at the "sausage" tyre. For the protagonists, the introduction of the air tyre was intended to solve the vibration problem of the solid tyre. However, vibration presented a serious problem to the low-wheeled bicycles. The high-wheeled bicycles did not suffer much from this problem. But what changed the situation was that when the low-wheelers began to accept the air tyre, they were able to outpace the high-wheelers. In bicycle racing, handicappers (i.e., high-wheelers and low-wheelers on solid tyres) had to be given a considerable start if riders on air tyres entered the race. After a while all concerned groups, racers, and ordinary users went for speed and stability and thus the closure was achieved by redefining the original problem. See Pinch and Bijker, "Social Construction," 44–45.

34. Bijker, *Sociotechnical Change.*

35. Bijker, *Sociotechnical Change,* 98–99.

36. Wiebe E. Bijker, "The Social Construction of Bakelite: Towards a Theory of Invention," in *The Social Construction of Technological Systems: New Directions in the Sociology and History of Technology,* Wiebe E. Bijker, Thomas P. Hughes and Trevor Pinch, eds. (Cambridge, MA: MIT Press, 1987), 159–187.

37. Pinch and Bijker, "Social Construction."

38. Bijker, "Theory of Invention," 172, emphasis in the original.

39. Bijker, "Theory of Invention," and *Sociotechnological Change.*

40. Bijker, *Sociotechnical Change,* 123.

41. Bijker, *Sociotechnical Change,* 123.

42. Electrical manufacturing companies, such as General Electric, Westinghouse, and Remy Electric, were Baekeland's (the inventor of Bakelite) first customers. They bought the molding material from Baekeland's company, General Bakelite Company. It was by means of the electrical industry that Baekeland was able to enroll the automobile industry, the largest consumer of Bakelite, which initially used Bakelite for mounting electrical parts, and later for making such nonelectrical parts as steering wheels, radiator caps, gear shift knobs, and door handles. See Bijker, "Theory of Invention," 177.

43. Bijker, *Sociotechnical Change,* 125.

44. Bijker, *Sociotechnical Change,* 104–105.

45. John Clark, Ian McLaughlin, Howard Rose, and Robin King, *The Process of*

Technological Change: New Technology and Social Choice in the Workplace (Cambridge: Cambridge University Press, 1988).

46. Conversion of network operations from an analog mode to digital mode is one of the most significant technological changes now taking place in the telecommunications field. Digital switching, digital transmission, and common control using digital computers and microprocessors, along with advances in satellites and fiber optical wave-guides, among others, is revolutionizing telecommunications. Though digital technology is not the empirical focus the authors discuss the "potential implications" of digital exchanges in one chapter at the end. Despite the dated nature of the empirical example, this is an interesting case study of technological change that can shed important light on the process of technological change.

47. For a critical review of this book, see Govindan Parayil, "Book Review: *The Process of Technological Change: New Technology and Social Choice in the Workplace* by John Clark, et al." *Science, Technology & Human Values* 15, no.1 (1990): 124–125.

48. Donald MacKenzie, "The Missile Accuracy: A Case Study in the Social Processes of Technological Change," in *The Social Construction of Technological Systems,* Wiebe E. Bijker, Thomas P. Hughes, and Trevor Pinch, eds. (Cambridge, MA: MIT Press, 1987), 195–222.

49. John Law, "The Anatomy of a Socio-Technical Struggle: The Design of the TSR 2," in *Technology and Social Process,* Brian Elliot, ed. (Edinburgh: Edinburgh University Press, 1988), 44–69, "On the Social Explanation of Technical Change: The Case of the Portuguese Maritime Expansion," *Technology and Culture* 28, no. 2 (1987): 227–252, and "Technology and Heterogeneous Engineering: The Case of Portuguese Expansion," in *The Social Construction of Technological Systems,* Wiebe E. Bijker, Thomas P. Hughes, and Trevor Pinch, eds. (Cambridge, MA: MIT Press, 1987), 111–134.

50. Thomas P. Hughes, *The Networks of Power: Electrification in Western Society, 1880–1930* (Baltimore: Johns Hopkins, 1983).

51. Inertial guidance or navigational control of vehicle—missile, aircraft, submarine, or spaceship—works by measurement of the acceleration experienced by the inertial measurement unit of the vehicle. The accelerometers in tandem with a set of gyroscopes, satellites, radar altimeters, feedback controls, and an onboard digital computer navigate the missiles. The onboard computer will have crucial data, such as a gravity map and geodetic data. On some missiles, the navigational system may be supplemented with the vehicle to sight particular stars, thus providing additional input as referents. For details, see MacKenzie, "Missile Accuracy."

52. Law, "Socio-Technical Struggle."

53. The case study was rewritten using the actor network model in line with the latest SCOT tradition of "interpretive flexibility" of artifacts as the theoretical frame by John Law and Michel Callon, "The Life and Death of an Aircraft: A Network Analysis of Technical Change," in *Shaping Technology/ Building Society,* Wiebe E. Bijker and John Law, eds. (Cambridge, MA: MIT Press, 1992), 21–52. However, here I will follow only the first version developed by John Law, "Socio-Technical Struggle."

54. Law considers the construction of the artifact as part of an elaborate process of system building.

55. Law, "Socio-Technical Struggle," 46.

56. In simple terms, this step appears to be a detailed planning to initiate the technological problem solving.

57. Law, "Socio-Technical Struggle," 46.

58. "Obduracy" appears to be a metaphor that Law uses for Hughes's "reverse salients" or Rosenberg's "technological bottlenecks."

59. In addition to design problems, the problems raised by Britain's ministry of defence, treasury department, and the navy, and such contingent factors as the change in government, caused the demise of this technological innovation.

60. Law, "Technical Change," and "Heterogeneous Engineering."

61. The first westerner to successfully reach India by rounding the Cape of Good Hope was Vasco da Gama, who anchored his ship at Calicut on May 20, 1498. Da Gama's negotiations with the Samorin of Calicut to enter into spice trade were a total disaster. However, da Gama returned to Calicut in 1502 with an armada of galleys equipped with guns and cannons, and bombarded the town of Calicut to force the Samorin into submission. Although da Gama faced stiff resistance from Samorin's army and navy, he moved to the south of Calicut and made deals with other regional kingdoms to set up a spice trade. The Portuguese stayed in India until they were ejected by the Indian army in 1962 from the state of Goa, which was still under Portuguese control even though the British army left India in 1947.

62. Law, "Heterogeneous Engineering," 115.

63. Law, "Heterogeneous Engineering," 129.

64. Although Law claims that the systems model of Thomas Hughes influences his heterogeneous engineering model, there seems to be no connection between these models, other than using historical case studies as empirical examples.

65. Michel Callon, "Society in the Making: The Study of Technology as a Tool for Sociological Analysis," in *The Social Construction of Technological Systems: New Directions in the Sociology and History of Technology,* Wiebe Bijker, Thomas Hughes, and Trevor Pinch, eds. (Cambridge, MA: MIT Press, 1987), 83–106.

66. Callon, "Society in the Making," 93.

67. Callon, "Society in the Making," 88, as paraphrased by Callon.

68. Bruno Latour, *Science in Action: How to Follow Scientists and Engineers Through Society* (Cambridge, MA: Harvard University Press, 1987).

69. Latour, *Science in Action,* 27.

70. One of the major contentious issues of the network model is that Latour, Callon, and others attribute agency and reflexivity to all elements of the actor-network regardless of their differing constitutions.

71. Latour, *Science in Action.*

72. For an expanded narrative of the Diesel story, see Donald E. Thomas, *Diesel: Technology and Society in Industrial Germany* (Tuscaloosa, AL: University of Alabama Press, 1987).

73. Wiebe E. Bijker, "The Social Construction of Fluorescent Lighting, or How an Artifact Was Invented in Its Diffusion Stage," in *Shaping Technology/Building Society: Studies in Sociotechnical Change,* Wiebe E. Bijker and John Law, eds. (Cambridge, MA: MIT Press, 1992), 75–104, and *Sociotechnical Change.*

74. Bijker, "Fluorescent Lighting," 76.

75. "Mazda" companies are commonly referred to as those corporations that used the incandescent lamp trademark "Mazda." General Electric and Westinghouse Corporations are the two important actors that Bijker follows in his example.

76. The precursor of the fluorescent lamp was the electric discharge lamp where the light generated by a gas discharge in a tube is used for illumination. The practical applications of such lamps were limited to outdoor lighting and displays. Interest in developing daylight discharge lamps led to the possibility of using fluorescent materials for getting light closest to "daylights." See Bijker, *Sociotechnical Change.*

77. Bijker, *Sociotechnical Change,* 217.

78. Bijker, "Fluorescent Lightning," and *Sociotechnical Change.*

79. Bijker, "Fluorescent Lightning," and *Sociotechnical Change.*

80. Bijker, *Sociotechnical Change,* 261.

81. Bijker, *Sociotechnical Change,* 263.

82. Bijker, *Sociotechnical Change,* 263–264.

83. Langdon Winner, "Upon Opening the Blackbox and Finding It Empty: Social Constructivism and the Philosophy of Technology," *Science, Technology & Human Values* 18 (1993): 362–378.

84. The radical constructivist accounts of Latour, Callon, Law, and others might find this accusation misplaced, for they refuse to treat the social as a distinct background on whose backdrop the social relevance has to be distinguished.

85. A caveat is in order in light of the technological frame model of technological change developed by Bijker (analyzed earlier), who incorporates the concept of power in narrating the social construction of artifacts. For an excellent response to the critiques of the SCOT program, particularly its alleged political vacuousness, see Trevor Pinch, "The Social Construction of Technology: A Review," in *Technological Change,* Robert Fox, ed. (Amsterdam: Harwood Academic Publishers, 1996), 17–35.

86. Steven Yearly, *Science, Technology, and Social Change* (London: Unwin Hyman, 1988).

87. Yearly, *Technology and Social Change,* 11.

88. Walt W. Rostow, *The Stages of Economic Growth: A Non-Communist Manifesto,* 3d ed. (Cambridge and New York: Cambridge University Press, 1990).

89. W. Arthur Lewis, "Economic Development with Unlimited Supplies of Labour," in *The Economics of Underdevelopment,* A. Agarwala and S. Singh, eds. (Oxford: Oxford University Press, 1954), 400–449.

90. Dependency theory originated in Latin America in the 1960s to explain the economic and political backwardness of Latin American countries and particularly in the wake of these countries' subservient relationship to American imperialism. Some of the founding theorists of dependency theory are Andre Gunder Frank, *Dependent Accumulation and Underdevelopment* (New York: Monthly Review Press, 1979), and Samir Amin, *Accumulation on a World Scale: A Critique of the Theory of Underdevelopment* (New York: Monthly Review Press, 1974). However, dependency theory as a coherent explanatory device for understanding the underdevelopment of the Third World is under retreat as admitted by one of its founding theorists, Gunder Frank himself. See Marta Fuentes and A. G. Frank, "Ten Theses on Social Movements," *World Development* 17, no.2 (1989): 179–191. For a comprehensive review of Yearley's book, see Govindan Parayil, "Yearley's *Science, Technology and Social Change:* Review," *Social Epistemology* 6, no.1 (1992): 57–63, and "Yearley's *Science, Technology and Social Change:* Reply," *Social Epistemology* 6, no.1 (1992): 73–75.

91. Yearly, *Technology and Social Change,* 150.

92. Wiebe E. Bijker and John Law, "General Introduction," in *Shaping Technology/Building Society: Studies in Sociotechnical Change,* Wiebe E. Bijker and John Law, eds. (Cambridge, MA: MIT Press, 1992), 1–16.

93. Bijker and Law, "General Introduction," 10.

94. Sismondo, "Some Social Constructions."

95. Bijker and Law, "General Introduction," 13.

96. Bruno Latour, *Aramis, or The Love of Technology* (Cambridge, MA: Harvard University Press, 1996).

4

Economic Models

ECONOMISTS AS PIONEERS

Despite being the pioneers in recognizing the importance of technological change as the key factor responsible for economic growth and social change, modern economists in general seem to have become stuck in the same mode of thinking and theorizing, unlike their predecessors, as evidenced by their narrow range of thinking and scholarship on explaining technological change. Despite being recognized as one of the pioneers in supposedly elucidating the role technological change plays in economic growth, Robert Solow does not show us what exactly constitutes technological change.[1] For all practical purposes, Solow and other neoclassical economists conceptualize technology as an exogenous factor that is outside the production process (function). This, notwithstanding the prime importance accorded in their canons to technological change as the causal factor that induces long-term growth and productivity increases. As we will show later, in fact, it was rather serendipitously that neoclassical economists came to know that it was technological change that caused almost all the increases in productivity growth. Although economics is not a unified discipline, because of the power neoclassical economics wields over this discipline, its ideas on technological innovation and change dominate the field.

Economists, particularly, in their political-economic garb had long discoursed on the importance of technology in the modernization project since the Enlightenment. The birth of industrial capitalism from the ruins of feudalism in Western Europe owes a great deal to the growth of technological knowledge that Europe experienced. Enlightenment *philosophes* like Condorcet, Turgot, Diderot, and D'Alembert and classical political economists like Smith, Ferguson, Petty, Babbage, Mill, and Marx recognized the importance of scientific and technological knowledge in the transition from traditional agrarian societies to modern industrial societies. Adam Smith explored the profitability of extending the age-old principle of the social division of labor on the shop floors of newly emerging factories so as to revolutionize manufacturing. In Volume I of *Capital,* Marx provides detailed descriptions of the historical development and change of industrial technologies and the impact of the division of labor on manufacture and the labor process. Marx provides an illuminating commentary on the nature of technological change in machinery and large-scale mechanized production, as we already saw in chapter two. In this chapter, however, we will look at how modern economists of various persuasions and genres conceptualize technological change. But before analyzing the current scholarship on the subject, let us briefly look at the classical economic approaches to understanding and explaining technological change.

CLASSICAL APPROACHES TO TECHNOLOGICAL CHANGE

Technological change is the most important conceptual problem for economists, whether they state it explicitly it or not in their theories and models. Classical political economists, such as Adam Ferguson,[2] Adam Smith,[3] David Hume,[4] and Karl Marx,[5] gave a prominent place in their writings to technological change, and argued that it is one of the most important causal forces responsible for the rise of manufacturing and industrial capitalism that led to lasting social and economic changes. Despite their intuitively strong intellectual stirring on the topic, these analysts, however, did not provide detailed theoretical accounts of how to conceptualize it, nor did they expound clear accounts of how to come to terms with this phenomenon.

Hume, Smith, and Marx were attempting to provide an account of how a predominantly agrarian (English) society changed into one where industry and commerce flourished and became the first industrialized country in the world. Hume's intellectual work, particularly those related to political economy, had a profound influence on Smith's theories of economics.[6] The closest that Smith came to expounding an intuitively interesting conceptualization of technological change in economic discourse was his theory of the division of labor.[7] Although the concept of division of labor based on different social categories (such as gender, class, caste, age, and so on) was as old as all human societies, Smith's exhortation of its application to the production processes in industrial settings gave great impetus to the nascent Industrial Revolution.[8] The revolutionary concept of the division of the labor process into specialized activities led not only to the increase in labor productivity, but also to the eventual development of specialized tools, machinery, and production systems.[9] However, as Ferguson and Marx observed, the dividing of labor time leads to the alienation of the laborers from the labor process, including the means of production, by making work monotonous, physically exhausting, and intellectually stultifying. However, Smith's ideas on the division of labor and the productivity puzzle were the ones that interested most economists.

After Smith, serious and original reflections on the economics of technological change originated with Marx. Marx's interest in technology may be elicited from his keen observations on the "progressive" nature of industrial capitalism stimulated by technological change. By progressive what Marx meant was that the forces of production of industrial capitalism would deal a deathblow to the archaic feudal relations of production, thus paving the way for the birth of the modern proletariat, which has the consciousness to understand the material conditions of their existence.[10] The bourgeoisie also realized that in a competitive local and global marketplace, it needs all the benefits of modern technology for its survival and reproduction. According to Marx (and his coworker Engels), "[T]he bourgeoisie cannot exist without constantly revolutionizing the instruments of production."[11] Unlike the feudal lords and the mercantilists, the bourgeoisie knew the importance of the new methods of industrial production and the urgent need for newer technologies, and considered it very important to invest in new machines that would create new machines.

The new instruments of production that have evolved as a result of technological innovation help the bourgeoisie to extract the maximum surplus from the labor, whose labor, according to Marx, is the only factor that creates wealth or adds value to a product during the production process. The resulting immiseration (according to Marx, the workers are paid a miserable subsistence wage), deskilling of the laborers, and their alienation from the means of production, not only due to the adverse relations of production but also due to the mind-numbing division of labor introduced in industrial production, would force the workers to rise up against the bourgeoisie and the state that helps to perpetuate this unequal exchange. Marx contends that there is a dialectical relationship between changes in the modes of production, which may be construed as technological change, and changes in the relations of production. However, Marx's prediction of the imminent overthrow of the capitalist system by violent insurrections of the proletariat did not take place in England and other rapidly industrializing capitalist economies of Western Europe. The radicalization of the working class through labor mobilizations and the inherent dynamism of capitalists to co-opt the working class and its interests through a shorter workweek, higher wages, and benefits may have staved off the predicted revolutionary overthrow of the capitalist system.

While fulminating against the inequities of the capitalist modes and relations of production in the industrializing colonialist powers like Britain, Holland, and Germany, Marx, however, saw the transfer of these technologies to the colonies as a blessing in disguise. It was hoped that these modern technologies, once transferred to the colonial outposts by the colonial powers, would undermine the feudal relations of productions existed in countries such as India, China, and Ireland. Marx's commentaries on the colonial question show his great hope that the transfer of such technologies as railways, electricity, steam engines, telegraph, and automated looms would render a resounding blow to the "Oriental Despotism" and other archaic and exploitative productive systems, thus building the material foundations for a modernized productive system. The material foundations of a modern industrial system was expected to instill the consciousness in the newly formed industrial proletariat to overthrow not only the colonial masters but the local feudal and religious masters as well. However, Marx's expectations did not materialize because the technological change and the ensuing industrialization drive introduced in the colonies by the colonialists were partial and unarticulated. Rather, the new technologies were transferred for normalizing, pacifying, and subjugating the native populations and their lands for large-scale transfer of wealth and raw materials to the colonizers' lands.

A major allegation made against Marx's conceptualization of technology was that he was a technological determinist.[12] That is, being an epiphenomenon of the economic base, Marx was supposed to believe that technological forces determine social change. Most scholars use Marx's aside remark (while discoursing on the "metaphysics of political economy") that "the hand-mill gives you society with the feudal lord; the steam-mill, society with the industrial capitalist"[13] to make this charge. Imputing to Marx the discovery that "machines make history," Robert Heilbroner claims that technological determinism is the hallmark of the Marxian paradigm.[14] Lewis Mumford reads Marx's analysis of technological change as one in which technological forces determine the character of all social institutions.[15] The most straight forward interpretation of the Marxian model of technological change as the primary determinant of social change was made by William Shaw, who read in Marx that the forces of production (technology) determine social change.[16]

Although the normative dimensions of the Marxian model is beyond the scope of discussion here, imputing technology as the independent variable in Marxian analysis of historical change (of socioeconomic systems) may be simplistic. Although statements taken out of context from Marx's prolific writings can be used to assert a claim of technological determinism, Marx was far too complex a scholar to be tagged with such an accusation. In fact, in Volume I of *Capital,* Marx extensively dealt with the issue of "the most significant complex of technical changes of his time, the coming of large-scale mechanized production," and held the view that "social relations molded technology, rather than vice versa."[17] Thus, according to Nathan Rosenberg, for Marx,

> Invention and innovation, no less than other socioeconomic activities, were best analyzed as social processes rather than as inspired flashes of individual genius. The focus of Marx's discussion of technological change is thus not upon individuals, however heroic, but upon a collective, social process in which the institutional and economic environments play major roles.[18]

Marx clearly delineated the fact that the issue is more about who is in control of technological innovations. In a capitalist system, it happens to be the owners of the means of production. In fact, the alienation of the proletariat from the means of production, pointed out by Marx, is a sign of the imminent doom of the capitalist system.[19] Marx was more than a mere economic interpreter of technological change, but a grand social theorist with strong historicist convictions who was well aware of the contingent nature of technological change.

Serious works on technological change from the economists' side declined after Marx, to be picked up again in the early part of this century by Joseph Schumpeter.[20] Schumpeter took keen interest in explaining the dynamics of economic development spurred on by technological change. He analyzed "economic evolution" using the unit of analysis of business cycles. According to him, "creative destruction," a process during which new technological developments evolve, is the fundamental basis of how capitalism works, and this process, again, is spearheaded by innovating entrepreneurs. In Schumpeter's model, a new technological innovation in the existing production process destroys the old process. Schumpeter considered only innovation as the most important structural component of technological change, and considered invention an act outside the realm of legitimate economic practices. Schumpeter considered innovation as the setting up of a new production function, and held that innovations cluster because of business cycles.[21] Schumpeter's assumption of "creative destruction" in which a new technology introduced under the leadership of innovating entrepreneurs always replacing old technologies is only partially true.[22] According to Strassmann, old technologies do often coexist in an industry, and total technological change (that is, the complete replacement of an old technology by a new one) takes place only when the "rate of obsolescence" occurs with "unforeseen rapidity."[23] Carolyn Solo questions the Schumpeterian stricture that invention is a practice outside legitimate economic practices.[24] She contends that industries spend money for R & D activities with specific objectives of inventing new technologies. Usher also criticizes Schumpeter for ignoring inventive activities in his analysis of technological change, and claims that Schumpeter's heroic innovator is similar to the inventor in most circumstances.[25] However, I will show in chapter five that Schumpeter's ideas on technological change are the most original, perhaps besides Marx's, that influenced a rich array of models and theories of technological change during the past several decades.

NEOCLASSICAL APPROACHES TO TECHNOLOGICAL CHANGE

The deliberate indifference on the part of neoclassical economists to the ideas on technological change and economic dynamics of Marx, Veblen, Schumpeter, and others became embarrassingly evident as they tried to grapple with the project of explaining and theorizing the process of economic growth and change without paying attention to historical hindsight and social factors. Despite an intuitive knowledge of how advances in technological knowledge change things in society (specifically, changes in the material conditions of living), often for the better, neoclassical economists could think of the effect of technological change only as an exogenous process, a "thing" hidden in the production process that they could conceptualize only in a roundabout manner. Technological change became, for them, a mysterious force that somehow helps the material conditions of society effected through raising the rates of economic growth and productivity increase.

Following Lakatos's methodology of scientific research program,[26] the "hard core" of the neoclassical program consists of three theses: (1) The thesis of equilibrium via the market mechanism;[27] (2) The theses of marginal utility, marginal productivity, and the method of comparative statics;[28] and (3) The "homo oeconomics" thesis that assumes "human beings as simply a bundle of preferences."[29] Colin Glass and W. Johnson put the above theses, alternatively, as individualism, rationality, private property rights, and market economy.[30] The "positive heuristics" of the neoclassical program, according to Blaug, are (1) divide the markets into buyers and sellers, or producers and consumers, (2) specify the market structure, (3) create "ideal type" definitions of the behavioral assumptions so as to get sharp results, (4) set out the relevant *ceteris paribus* conditions, and (5) translate the situation into an extreme problem and examine first- and second-order conditions.[31] The "protective belt" or the auxiliary assumptions are: (1) rational economic calculations, (2) constant tastes, (3) independence of decision making, (4) perfect knowledge (of market conditions), and (5) perfect mobility of factors of production.[32]

The typical methodology adopted by neoclassical economists to analyze economic growth and its linkage to technology is to use an aggregate production function (popularized by Solow in his 1957 classic paper),[33] which is based on the assumption that one can develop an aggregate production function for an industry or nation by aggregating the output of all firms as well as the total workforce and invested capital. That is, combine the individual production functions of all the firms and other units of production in the economy. A production function is typically an input-output relationship. A production function specifies the relationship between various combinations of inputs (or factors of production) and outputs, which can be represented algebraically as akin to an engineering production function that can also be graphically represented by smooth, convex isoquants representing different methods that can be adopted in the production of a given volume or amount of output. Technological change is represented by the shift of the isoquant.[34]

A model of a typical neoclassical (aggregate) production function can be represented as: $Q = f(K,L;t)$, where 'Q' is the aggregate output, 'K' the total capital stock, 'L' the total number of workers in the national labor force, and 't' is time.[35] The share of output to capital and labor may be ascertained by assuming a Cobb-Douglas production function, which may be represented as in equation (1):

$$Q = f(K^a, L^b) \tag{1}$$

where coefficients a (increase in productivity attributable to advances in capital stock or capital's share in national income) and b (increase in productivity due to the addition of labor in the production process or labor's share in national income) can be calculated by regressing time series value data, on Q, K, and L, available from national income accounts. The sum of the coefficients ($a+b = 1$) should be the total increase in production.[36] The above method will provide only the instantaneous ($t = 0$) productivity details. The total factor productivity over a period t

requires a slight modification of the production function, (which is what Robert Solow did in his famous 1957 growth study model), as we will see below.

It is quite clear that the above neoclassical model of an economy with only two variables as input factors is rather simplistic and ahistorical, but appears elegant for presenting a simplified picture of market economies. This becomes obvious when neoclassical economists "discovered" technological change as the predominant factor that caused economic growth. In a growth study of the United States economy from 1869 to 1953, Moses Abramovitz came across a great puzzle.[37] After attributing increase in net national product to the two input variables, capital and labor, Abramovitz could not account for the factor that was responsible for the largest increase in net national product. Abramovitz in great frustration for not being able to identify this mysterious factor avers,

> [S]ince we know little about the causes of productivity increase, the indicated importance of this element may be taken to be some sort of *measure of our ignorance* about the cause of economic growth in the United States and some sort of indication of where we need to concentrate our attention.[38]

However, Abramovitz alluded that this mysterious factor is technological change. It is clear that the measure of neoclassical economists' ignorance is precisely due to their ignoring the works of Schumpeter, Marx, and others.

While Abramovitz claimed that economists were ignorant of how the largest share of economic growth can be explained, Robert Solow, while analyzing the growth of the United States economy (nonfarm sector) from 1909 to 1949, boldly conjectured that 90 percent of the increase in gross output per man hour should be attributed to technological change without providing convincing evidence, while he accounted for the remaining increase in output to capital deepening.[39] Thus this mysterious "residual" factor of Abramovitz metamorphosed into technological change in Solow's interpretation. According to Lester Lave, it was Evsey Domar who appropriately dubbed this factor the "residual," that is, that segment of the increased output per man hour that is left over after increases in capital per labor are accounted for.[40] According to Lave, the method by which Solow measured the "residual" (the so-called technological change index) is: Solow assumed an aggregate production function of two factors, capital and labor, that is characterized by constant returns to scale.[41] This aggregate production function can be represented as:

$$Q = A(t)f\{K(t), L(t)\} \qquad (2)$$

'A' in equation (2) is supposed to represent the state of technology in the nation state or social unit of analysis. Solow also assumed that the economy is perfectly competitive, the production function is homogeneous, and the technological change is neutral (that is, isoquants shift out parallel to each other). Therefore, the technological change between two time periods can be deduced from the equation:

$$\Delta A/A = \Delta Q/Q - w_k\Delta K/K - w_l\Delta L/L \qquad (3)$$

where, w_k and w_l are the elasticities of output with respect to capital and labor. In the above equation, technological change is equal to the change in the output not accounted for by the change in output due to capital and labor. Solow's simplified model fitted to a Cobb-Douglas production function can be stated as:

$$Q = A.f(K^a,L^b) \qquad (4)$$

where A is identified as the factor representing the state of technological change, and b $= (1-a)$.[42] It is interesting to note that A was not placed inside the bracket, because neoclassical theory recognizes only capital and labor as input variables. It became an exogenous variable, and hence the well-placed criticism that neoclassical economists "black box" technological change. In time t, the production function can be written as:

$$Q_t = Ae^{ct}K_t^{a}L_t^{1-a} \qquad (5)$$

The coefficient 'a' can be measured by ascertaining capital's (K) share in national income. Solow's accounting method had some curious results. He found out that the increase in productivity attributable to capital and labor during the period of his study came to only 0.125 or 12.5 percent of productivity increase. The remaining productivity increase (0.875 or 87.5 percent) had to come from somewhere else. By taking the derivative of equation (5) with respect to time, and using the readily available data on Q, K, and L from national income accounts, one can solve for 'c', the average increase in total productivity.[43] Solow estimated 'c' to be about 1.5 percent per annum, which amounts to 87.5 percent for the entire duration of his study. This "residual" is what Solow identified as technological change. Thus, Solow's model, to a large extent, set off a thriving research program in neoclassical economics during the past forty years to find the elusive "residual" using the production function model of technological change.[44]

Serious criticisms of the production function model of technological change came from quarters within neoclassical economics (economists opposed to the production function) and from Keynesian and Marxian economists. Richard Nelson and Sydney Winter, from within the neoclassical ranks, offered one of the most devastating criticism of the residual approach to accounting for technological change as empirically spurious and an "intellectual sleight of hand."[45] They claim that the production function theory cannot model, and hence cannot represent technological change. Instead, Nelson and Winter argue, the residualists cover up the inadequacy by resorting to spurious empiricism. According to this view, labelling technological change as a residual is similar to the "neutrino" affair in physics, a case of labelling an erroneous discovery that later proved to be useful. This is evident from the fact that despite any theoretical knowledge of how technological change creates economic growth and wealth, the unprecedented enthusiasm shown by the growth theorists unwittingly resulted in policies such as investing in R & D to spur on technological innovation.

Besides treating technological change as an exogenous variable, the production function approach ignores the temporal and cumulative nature of this essentially continuous and qualitative phenomenon. The production function model assumes that the actors or the units of analysis of the economic process, which are mostly firms, opt for a new technology under the postulates of perfect competition, perfect knowledge, and the easy availability of a range of "alternative" technical solutions from a "technology shelf" for solving the same production problem. That is why Nathan Rosenberg argues that since the development of a new technology itself is a costly activity, alternative technological solutions representing factor combinations far from those justified by current factor price levels need not be known in advance to allow a shift of the production isoquant in the first place.[46] That is, if superior technological alternatives were known in advance, why would a prospective entrepreneur choose the inferior technology in the beginning of the production process? The production function model does not shed any light on how and in what sense is it likely to be the case to have knowledge of technological alternatives, given the fact that technological knowledge production is a costly activity. Being grounded in the tenets of static equilibrium, the atemporal production function model does not help itself to learn from historical agency and experience in economic analyses.

A serious flaw of the production function model is the assumption of a "logically pure" production function that "represents all conceivable techniques of production that *could* be designed by means of the existing knowledge."[47] This again comes from the presumption that one could design various tools and production methods based on every known physical and biological law. As Sahal accurately points out, this notion hinges on the belief that technology originates from pure science. That is, technology is applied science, pure and simple. But we know that despite our advances in understanding the intricacies of nature and natural processes, we cannot design technological devices to conform to such an understanding. What may be theoretically possible is not practically feasible, because of numerous constraints, economics being a prime one. One can think of a theoretical production function, say the generation and transmission of electric power, based on the knowledge of superconductivity. However, this is useless unless we can find economically feasible high-temperature superconducting materials that can be developed for building large turbine generators, transformers, and power transmission cables. Also, it is crucial to obtain the conditions for stimulating the technological change that can make this technology a reality.

The most damning criticism against the neoclassical production function is about how realistic is the possibility of constructing an aggregate production function, and how feasible it is to empirically test it using a neoclassical production function model such as Cobb-Douglas. Both the Solow (exogenous) and the new

(endogenous) neoclassical growth theories take the viability of an aggregate production function as axiomatic. Production function being a concept that is valid,
for the most part, at the microlevel (firms and other actors), aggregating these for
the whole economic system and then representing the aggregated economic system by the same firm-level production function is highly questionable. The aggregation problem raises serious problems like measuring the factors of production,[48] whether value data in monetary terms should be used or data in physical
terms such as work and energy units. How is it possible for one to combine different industrial enterprises like automobiles, textile manufacturing, coal mining,
banking, and so on into a single, representative aggregate production function?
Alan Walters puts the problem of aggregation more poignantly:

> After surveying the problems of aggregation one may easily doubt whether there is
> much point in employing such a concept as an aggregate production function. The va-
> riety of competitive and technological conditions we find in modern economies sug-
> gest that we cannot approximate the basic requirements of sensible aggregation ex-
> cept, perhaps, over firms in the same industry or for narrow sections of the economy.[49]

However, despite raising apprehensions about the usefulness of an aggregate production function, Walters and others still use it for testing of economic growth
trends using value data drawn from national accounting statistics.

The validity of the estimations of aggregate production functions using the
Cobb-Douglas formula was seriously challenged by Anwar Shaikh and Herbert
Simon.[50] Shaikh's powerful critique may be put simply that the Cobb-Douglas
production function using value data cannot be empirically tested because the regression estimates are conducted on an accounting identity, specifically, a cost
identity of the form: $Q = rK + wL$, where Q, K, L, r and w are, respectively, total output (value added during the production process), capital stock, employment
level, rate of profit, and wage rate. Shaikh argues that the production function
Solow used was a "humbug" production function because it can be derived from
the above cost identity, which can be easily proved by the following simple steps.
The total output at time t can be written as:

$$Q_t = r_t K_t + w_t L_t \tag{6}$$

Differentiating equation (6) with respect to time, we obtain the instantaneous rates:

$$q_t = c_t + a_t k_t + (1-a_t)l_t \tag{7}$$

where
$$c_t = a_t c_{rt} + (1-a_t)c_{wt} \tag{8}$$

The variables q_t, k_t, and l_t denote growth rates of output, capital, and labor; and c_{rt}
and c_{wt} are the rental price of capital and growth rates of wages, and a_t is capital's

share in total output (that is, $a_t = r_t K_t / Q_t$ and $[1-a_t]$ is labor's share in total output). Assuming that these shares are constant over time and integrating equation (7) we obtain:

$$Q_t = Ae^{ct}K_t^a L_t^{1-a} \tag{9}$$

Surprisingly, the above equation (9) derived from the cost accounting identity (6) is the same as the Cobb-Douglas production function Solow had used earlier (5) to estimate the factor shares and the "residual" technological change index. Thus, by actually regressing an accounting identity (cost function), the production functions give very close fit to the data. These devastating findings have not shaken the faith in the neoclassical production function, however, as evidenced by the continuing proliferation of growth accounting studies to find out the role of technological change in economic development.[51]

In the production function model, the emphasis has always been on a rational-actor approach that attempts to explain technological change as a condition to be achieved by choosing the "right" mode of production, or the most "optimal" technologies for solving an economic problem from an imaginary "technology shelf." Being based on using discrete value data by reducing entities of heterogeneous physical attributes to money or some other taxonomic practice, neoclassical theory fails to see technological change in a transformational sense of reordering nature and human-made structures and institutions. Since the basis of production function analysis is to use discrete aggregates on an instantaneous basis, it fails to comprehend the functional-cum-intentional nature of technological change.

Serious efforts to save the residual empiricism promoted by Solow and others came recently from the growth accounting studies of Lau, Kim and Lau, and Krugman, among others who are also adherents of the neoclassical production function paradigm.[52] However, more than saving the residual technological growth accounting practice, the works of these analysts may, in fact, unwittingly destroy the neoclassical model. While analyzing the fast economic growth achieved by many southeast Asian states, these economists came up with a puzzle. They came to the conclusion that the growth these East Asian "tigers" achieved came from input accumulation rather than from technological change, because the residual, according to their calculations, turned out to be negligible or even negative, despite the fact that the methodology used in these studies was the same growth accounting in neoclassical economics where factors are paid their marginal productivity shares.[53]

Because of these and other anomalies, neoclassical economists like Paul Romer and Robert Lucas attempt to endogenize the residual (technological) factor within the production function.[54] The attempt can be equated to an effort to open the black box of technological change (represented by A in Solow's model above) and represent it as input factors within the bracket through various parameters such as human capital, education, and so on as proxy variables for technological change.

However, despite the novelty of the approach, Romer and others are actually trying to put old wine into a new bottle. They still use the neoclassical production function model to account for technological change. Romer attempts to decompose the
technology index A into variables like education 'E', human capital 'H', and so on,
and transfer these variables within the production function itself. Consequently, the
new improved "endogenous" production function would be as follows:

$$Q_t = Ae^{ct}(K_t^a, L_t^b, E_t^x, H_t^y, ...) \tag{10}$$

As more variables for the effect of technological change such as 'E', 'H', and so
on are introduced, the residual coefficient 'c' becomes smaller and smaller. It captures the effect of technological change not yet captured by the endogenized variables in the production function. The coefficients, x for increase in productivity
due to education, y for increase in productivity due to human capital, and so on can
be ascertained in the same way as the capital and labor coefficients are computed
in equation (5).

Romer provides the concept of "mixing" in chemistry as the appropriate
metaphor for technological development in economic activities.[55] Thus, the
knowledge of developing the most efficient mixing can be deemed as technological change. Certainly, a great improvement over the crude residual empiricism of
Solow and others; still resorting to the production function theory and other neoclassical orthodoxy will not save these "endogenous" models.

Endogenous explanation of technological change without resorting to a production function is attempted by Jacob Schmookler, who argues that economic
growth is caused by an essentially economically driven activity known as "invention."[56] Schmookler claims, based on his research on the American railroad industry, that inventive activity can be seen as directly correlated to increased sales
of capital rail equipment. Schmookler's demand-push model of technological
change ignores supply-side factors in inducing economic change, which Rosenberg sees as tautological because a case can be made to fit every historical account
of technological change to only demand factors.[57] A similar nonproduction function approach was adopted by Zvi Griliches, who claims that increase in agricultural productivity in the United States during the period from 1940 to 1960 can be
attributed to the improved quality of input factors such as better technical training
to workers, better seeds, and equipment.[58] Griliches's study of the diffusion of hybrid corn reveals that technological change is affected by the speed with which the
new technology is accepted by the farmers and the rate of profitability enjoyed by
them. However, Griliches's analysis also falls short because of his tendency to
privilege innovation that is only one structural component of technological
change.[59] Schmookler's and Griliches's analyses of technological change are theoretically closer to the institutional approach described in the next section than the
production function approach.

INSTITUTIONAL APPROACH TO TECHNOLOGICAL CHANGE

An institutional economics-based conceptualization of technological change called "theory of induced innovation"[60] was offered by Vernon Ruttan, Hans Binswanger, and Yujiro Hayami.[61] However, they retain the assumptions such as perfect competition and static equilibrium of neoclassical theory. Nevertheless, Binswanger and Ruttan differ with the laissez-faire attitude of the mainstream neoclassicists by arguing that:

> The demonstration that technical change can be treated as endogenous to the development process does not imply that the progress of either agricultural or industrial technology can be left to an "invisible hand" that directs technology along an "efficient" path determined by "original" resource endowments or by the growth of demand.[62]

The main thrust of the induced innovation model is that progressive institutional innovation and development directly cause the production of new knowledge that leads to technological change. Specifically, Binswanger refers technological change to the "result of the application of new knowledge of scientific, engineering, or agronomic principles to techniques of production across a broad spectrum of economic activity."[63]

Unlike Veblen and other more radical institutionalists who believed that technology is the dynamic factor in economic and social change, the neoclassical institutionalists hold that institutions are the dynamic factor. Hence, they argue that "creating a demand" for institutional change will take care of changes in the modes and means of the productive systems of modern societies. Technological change, they contend, would follow if individuals and groups of individuals believe that the "transaction costs" of bringing about institutional change are less, and that it is profitable to "design" new institutions with such attributes.

The induced innovation approach to technological change clearly sees the limits of markets as the only allocative mechanism for research and development activities leading to the development of new technologies. The clearest indication of this model's usefulness, in comparison with the production function model, in understanding technological change is its conceptualization of the output of research and development activities as information, "in the form of blueprints, chemical formulae, new seed varieties, and scientific or technical laws."[64] Information (knowledge) being a public good, it may not always be "profitable" for firms to invest in research and development activities that generate information, which usually ends up as a "free" good as well. Hence, governments and other public agencies have to either subsidize research and development activities that lead to information (knowledge) creation, or to conduct these activities themselves. This situation makes institutionalists sanguine to the possibility of formulating technology policy at the governmental level to stimulate technological change, unlike their neoclassical counterparts who see markets and private capital as the only mechanism for inducing technological change.

Also, institutionalists provide better explanations for technology transfer as a dynamic, and often a transnational process that involves such macro agents as nation states, multinational corporations, international aid agencies, research institutes, and so forth. Ruttan and Hayami distinguish three phases of international technology transfer: (1) material transfer, (2) design transfer, and (3) capacity transfer.[65] The transfer of new materials such as seeds, plants, animals, and machines characterizes material transfer. In order to make use of these materials in a better way systematic adaptive research is conducted either by the transferors or by the receiving governments and their agents. Design transfer involves the export of designs such as blueprints, books, and machines. The transferee tries to duplicate the designs transferred, sometimes with modifications and adaptations to suit the local conditions.

Finally, capacity transfer involves the transfer of "scientific knowledge and capacity."[66] The primary objective is to "create the capacity for the production of locally adapted technology according to the prototype technology existing abroad."[67] It often involves experts from the transferring country or agency travelling to the recipient countries or companies to train the "local" scientists and engineers to create the capacity to help themselves to conduct research and solve problems. The Green Revolution case study is a good example of this model. The development of a highly successful indigenous agricultural research capability in India was a capacity transfer (see chapter six, eight and nine for more details).[68] By focusing on the contexts and conditions of technological knowledge production as an integral part of the technology transfer process, induced institutional innovation models provide a better understanding of technological change in the field of economics.

However, by treating technological change as an essential by-product of institutional change, the induced innovation model does not present a clear picture of how technological change actually takes place. It assumes that by undertaking institutional change that would facilitate economies of scale and reduced transaction costs, the concerned individuals and groups would opt for that particular technology that would help them attain these goals. Although the institutionalists claim that institutional change is endogenous and that technological change is endogenous to institutions, this obvious circularity would make it difficult to explain how technological change actually takes place. Ultimately, such an outcome is tantamount to "black boxing" this phenomenon. The institutionalists' tendency to concentrate on secondary institutions and their neglect of culture, ideology, and class relations further weaken the relevancy of the induced innovation model's ability to explain technological change. In sharp contrast to the institutional and neoclassical economic approaches to technological change analyzed in this and the previous sections, the economic models presented in the next section rely on historical insights and contingencies as powerful explanatory devices.

PATH DEPENDENCE AND TECHNOLOGICAL CHANGE

Diverging sharply from the neoclassical and institutional approaches, Nathan Rosenberg and Paul David propose that the best way to conceptualize technological change is to adopt a path dependent analysis.[69] That is, economists must discard the atemporal production function framework to explaining technological change, and instead should pay attention to the social, political, and historical contexts of invention, innovation, and other technological development activities. According to Rosenberg, technological change can only be explained by paying attention to the behaviors and problem-solving procedures of economic agents. In order to conceptualize technological change, he argues that the analyst must shift his focus from highly aggregated to highly disaggregated mode of analysis. Rosenberg argues that "complex technologies create internal compulsions and pressures which, in turn, initiate exploratory activity in particular directions."[70] Thus, path dependence of technological change means that:

> One cannot demonstrate the direction or path in the growth of technological knowledge merely by reference to certain initial conditions. Rather, the most probable directions for future growth in knowledge can only be understood within the context of the particular sequence of events which constitutes the history of the system.[71]

In order to better understand and model technological change, one must have a clear grasp of the manner in which the production and the subsequent growth of knowledge took place. In the particular case in point, one must do it by taking into consideration the vital social, economic, and historical factors. As past analysis points to a particular trajectory of a technology's change and development, it may be possible to exploit that trajectory for further improvement based on the stock of existing knowledge. For example, Rosenberg points out the case of transistor that set the stage for exploiting a particular trajectory of miniaturization of components in the electronics and computer technologies.[72] Thus the modification and improvement of a subsystem or component initiates a process of change, culminating in further innovations and eventually to new technological systems. This certainly is a refreshing change from the production function account of technological change.

Paul David provides one of the best examples that can illustrate the need to incorporate "historical accidents" and contingencies in the economics of technological change. His story of the economics of "QWERTY" shows clearly the meaninglessness of adhering to rational choice as the underlying logic of economic transactions in technology choice.[73] David posits a very puzzling question. Why is the layout of the keyboards of all personal computers and typewriters (assuming English and other western European languages as the medium of discourse) the way it is? What caused the keyboard layout of QWERTYUIOP (the letters on the keyboard layout below the numerals) to be the "universal" standard? Despite the inventions and development of alternative keyboard arrangements that could supersede the QWERTY keyboard in speed typing, why did these better alternatives not survive? For instance, the DSK or Dvorak Simplified Keyboard system lets one type 20–40 percent faster than the QWERTY model.[74]

One of the major "historical accidents" in the QWERTY keyboard saga is that the practice of touch-typing was practiced and perfected in the 1880s on the Remington typewriter that had the QWERTY layout.[75] Because the knowledge and practice generated from a previous technological artifact earned a niche in the selection process, it became harder for competitors to get a foothold in the typewriter market. David offers three reasons for the QWERTY arrangement to have become "locked in" in the technological development and selection process: (1) technical interrelatedness, (2) economies of scale, and (3) quasi-irreversibility of investment.[76]

Technical interrelatedness means that the potential customers of the typewriters (hardware), who were mainly businesses, sought out secretaries who already possessed the "software" skills of touch-typing. The existing knowledge base of touch-typing was centered on the Remington QWERTY system that became entrenched by the time alternative systems evolved. The employers of these alternative "hardware" had to offer special incentives to retrain their employees who were already familiar with the QWERTY layout. Consequently, typists had fewer incentives to learn touch-typing on alternative systems:

> [T]o the degree to which this increased the likelihood that subsequent typists would choose to learn QWERTY, in preference to another method for which the stock of compatible hardware would not be so large, the overall user costs of a typewriting system based upon QWERTY (or any specific keyboard) would tend to *decrease* as it gained in acceptance relative to other systems.[77]

This is what is meant by system scale economies or decreasing user cost per unit. The system embeddedness of the QWERTY system means that the software component became a predominant economic consideration in the newly emerging office technology business. As a result, switching to a nonQWERTY system became economically infeasible for both typewriter manufacturers and office managers, leading to a situation of quasi-irreversibility of investments. The economics of QWERTY shows the importance of path-dependent analysis in understanding technological change.

CONCLUSION

There is no unanimity among economists as to what constitutes technology and how to understand and model technological change. At one end of the continuum technology is conceptualized in its purely material manifestation, while at the other end it is looked at as useful practical knowledge gained through research and development activities by firms and public institutions. The conceptualization of technological change using an aggregate production function in the neoclassical tradition, which is the predominant paradigm in economics, though illuminating in many respects, needs serious reformulation.

In addition to the controversy regarding the possibility of abstracting an aggregate production function for a whole economic system (of a nation or the world), the applicability of a Cobb-Douglas production function to represent the economic system has come under great scrutiny as the works of Shaikh, Simon, Felipe, and others show.[78] As McCombie astutely points out, the ability of the Solow model to be the reigning paradigm in neoclassical economics, despite the serious anomalies pointed out by Shaikh, Simon, and others, is attributable to the rhetorical strategy adopted by the followers of that paradigm.[79] The inevitable intellectual revolution to overthrow this entrenched paradigm has to wait. It appears that Max Planck's sad observation, pointed out by Thomas Kuhn, may be relevant in economics in that, "a new scientific truth does not triumph by convincing its opponents and making them see light, but rather because its opponents eventually die, and a new generation grows up that is familiar with it."[80] As Kuhn pointed out, after all, Joseph Priestly held on to his phlogiston theory of combustion and never accepted the oxidation theory of combustion, Kelvin never accepted the electromagnetic theory, and Copernicanism (heliocentric theory) found few followers until almost a century after Copernicus's death. Neoclassical economists seem to be stuck in a time warp, unaware of the changes that had taken place in the physics of the nineteenth century, which was the model that the founding fathers of neoclassical economics emulated. The reigning paradigms of nineteenth-century Newtonian physics (neoclassical economists' model) had been changed to the Einsteinian physics of the special theory of relativity in the early part of this century, which again was replaced by quantum theory a few decades later. Members of this supposedly scientific community seem to be unaware of history's powerful message that those who refrain from learning from it are condemned to repeat their mistakes until they become irrelevant.

ENDNOTES

1. Robert Solow, "Technical Change and the Aggregate Production Function," *Review of Economics and Statistics* 39 (1957): 312–320, and "Investments and Technical Progress," in *Mathematical Methods in the Social Sciences, 1959,* Kenneth J. Arrow, Samuel Karlin, and Patrick Suppes, eds. (Stanford, CA: Stanford University Press, 1960), 89–104.

2. Adam Ferguson, *Essays in the History of Civil Society* (New York: Garland Publishing, 1971).

3. Adam Smith, *An Inquiry into the Nature and Causes of the Wealth of Nations* (Chicago: University of Chicago Press, 1976).

4. David Hume, *Writings on Economics,* ed. and intro. by E. Rotwein (Madison: University of Wisconsin Press, 1955).

5. Karl Marx, *Capital: A Critique of Political Economy,* Volume I, tr. Ben Fowkes (New York: Vintage Books, 1977), and Karl Marx and Friedrich Engels, *Manifesto of the Communist Party* (Chicago: Encyclopaedia Britannica, 1952).

6. Martin Bronfenbrenner, "The 'Structure of Revolutions' in Economic Thought," *History of Political Economy* 3, no. 1 (1971): 136–151, 138, argues that Hume's political and economic discourses "anticipated Smith in so many matters of both positive and normative economics."

7. On the theory of the division of labor, Smith depended heavily on the writings of his teacher Adam Ferguson. In fact, Marx accuses Smith of blindly following, if not plagiarizing, Ferguson's writings on the division of labor.

8. Smith cites the famous pin factory example as the best application of the concept of division of labor in improving industrial productivity.

9. The next major ideas that revolutionized industrial production were F. W. Taylor's time and motion study–based theory of scientific management and Henry Ford's assembly line production system at his Detroit automobile factories. The synthesis of Taylor's and Ford's ideas of modern production found favor with not only the capitalists, but also with the socialist reconstruction tsars of Soviet Russia, particularly, Lenin.

10. The theory is that peasants and feudal serfs because they were mired in superstitions and metaphysics were wallowing in false consciousness, and hence could not understand the material conditions of their existence. That is, they did not understand how much they are being exploited by the feudal lords and religious leaders. An explicitly essentialist account of how only the proletariat knows the truth of the real nature of social relations of production because of its real life experience-based historical consciousness is elaborated by Georg Lukacs, *History and Class Consciousness: Studies in Marxist Dialectics* (Cambridge, MA: MIT Press, 1971).

11. Marx and Engels, *Manifesto,* 421.

12. Analysts who accuse Marx of being a technological determinist unreflectively quote him from the *Poverty of Philosophy:* "the hand-mill gives you society with the feudal lord, the steam-mill, society with the industrial capitalist."

13. Marx, *Poverty of Philosophy,* 95.

14. Robert L. Heilbroner, "Do Machines Make History?" *Technology and Culture* 8 (1967): 335–345.

15. Lewis Mumford, *The Myth of the Machine: Technics and Human Development* (New York: Harcourt Brace Jovanovich, 1966).

16. William H. Shaw, " 'The hand-mill gives you the feudal lord': Marx's technological determinism," *History and Theory* 18 (1979): 155–166.

17. Donald MacKenzie, "Marx and the Machine," *Technology and Culture* 25 (1984): 473–502, 473.

18. Nathan Rosenberg, *Inside the Black Box: Technology and Economics* (Cambridge: Cambridge University Press), 35.

19. For detailed accounts of denying the technological determinist label to Marx, see Mackenzie, "Marx and the Machine," Rosenberg, *Inside the Black Box,* and Richard W. Miller, *Analyzing Marx: Morality, Power, and History* (Princeton, NJ: Princeton University Press, 1984).

20. Joseph Schumpeter, *Business Cycles,* 2 volumes (New York: McGraw-Hill, 1939), and *Capitalism, Socialism, and Democracy* (New York: Harper & Row, 1950).

21. Schumpeter, *Business Cycles,* 84. More on production functions below. Schum-

peter worked within the classical (input factors: land, labor, and capital) and neoclassical (input factors: capital and labor) traditions. But most mainstream economists, particularly the neoclassical economists, did not take Schumpeter's works seriously. For more on this indifference on the part of the economists, see Robert J. Wolfson, "The Economic Dynamics of Schumpeter," *Economic Development and Cultural Change* 7 (1958/1959): 31–54.

22. See W. Paul Strassmann, "Creative Destruction and Partial Obsolescence in American Economic Development," *Journal of Economic History* 14, no. 3 (1959): 335–349.

23. Strassmann, "Creative Destruction," 336.

24. Carolyn S. Solo, "Innovation in the Capitalist Process: A Critique of the Schumpeterian Theory," *Quarterly Journal of Economics* 65, no. 3 (1951): 417–428.

25. Abbot P. Usher, "Technical Change and Capital Formation," in *The Economics of Technological Change: Selected Readings,* Nathan Rosenberg, ed. (Harmondsworth, UK: Penguin Books, 1971), 43–72.

26. Imre Lakatos, *The Methodology of Scientific Research Programmes* (Cambridge: Cambridge University Press, 1978).

27. Alfred W. Coats, "Is There a 'Structure of Scientific Revolutions' in Economics?" *Kyklos* 22, no. 2 (1969): 289–298.

28. Mark Blaug, "Kuhn Versus Lakatos, or Paradigms Versus Research Programmes in the History of Economics," in *Paradigms and Revolutions,* Gary Gutting, ed. (South Bend, IN: University of Notre Dame Press, 1980), 137–159.

29. B. Van Weigel, "The Basic Needs Approach: Overcoming the Poverty of Homo Oeconomics," *World Development* 14, no. 2 (1986): 1423–1434.

30. Colin J. Glass and W. Johnson, "Metaphysics, MSRP, and Economics," *British Journal for the Philosophy of Science* 39 (1988): 313–329.

31. Blaug, "Kuhn Versus Lakatos," 148. The first two of these five are generally undertaken to create "perfectly competitive" market structures to enable easy entry and exit for firms and other economic actors. The rest may be needed for optimization of resource allocation to achieve least cost production conditions and hence maximum profit.

32. Blaug, "Kuhn Versus Lakatos," 148.

33. Solow, "Technical Change."

34. An isoquant is defined as the locus of factor combinations that give the same output on a plot with labor on the horizontal axis and capital on the vertical axis. Alternatively, the horizontal axis could be capital per labor and the vertical axis could be output per labor. In the former case, the isoquant would be a convex parabolic curve, and in the latter case the isoquant would be an exponential curve. For the same output level with less of the input factors, if the isoquant shifts towards the origin in the former plot, economists say, there was a progressive technological change, and the curve shifts outwards in the latter case.

35. A major assumption is that factors "behave" according to constant return to scale, and capital and labor are paid their marginal products. Increasing or decreasing to scale would complicate the linear production function.

36. An alternative representation of the Cobb-Douglas production is: $Q = F(K^a, L^{1-a})$.

37. Moses Abramovitz, "Resource and Output Trends in the United States since 1870," in *The Economics of Technological Change: Selected Readings,* Nathan Rosenberg, ed. (Harmondsworth, UK: Penguin Books, 1971), 320–343. (This paper was originally published in 1956).

38. Abramovitz, "Resource and Output," 328, emphasis added.

39. Solow, "Technical Change."

40. Lester B. Lave, *Technological Change: Its Conception and Measurement* (Englewood Cliffs, NJ: Prentice-Hall, 1966), provides an excellent discussion and overview of the conception and measurement of technological change in neoclassical economics.

41. Lave, *Technological Change,* 11–14.

42. It is important to note here that as opposed to factor substitution (a movement along an existing portion of the production isoquant in response to a change in relative factor prices), technological change occurs, according to neoclassical theory, when the isoquant itself shifts resulting in an increase in the output for the same input.

43. See Paul Stoneman, *The Economic Analysis of Technology Policy* (Oxford: Clarendon University Press, 1987), for a simplified version of the method of finding the residual "c."

44. As Robert Solow, "Technical Progress," readily admits, the idea that the largest part of the increase in output per head was caused by "technical progress" was propounded by many other workers besides him, such as Abramovitz, "Output Trends"; Solomon Fabricant, *Basic Facts on Productivity Change* (New York: National Bureau of Economic Research, 1959); Jacob Schmookler, "The Changing Efficiency of the American Economy, 1869–1938," *Review of Economics and Statistics* 34 (1952): 214–231; and Theodore W. Schultz, "Reflections on Agricultural Production, Output and Supply," *Journal of Farm Economics* 38 (1956): 748–762. See Jesus Felipe, *Total Factor Productivity Growth in East Asia: A Critical Survey* (Manila, Philippines: Asian Development Bank, EDRC Report Series No. 65, 1997), for an interesting discussion of the proliferation of growth studies since Solow's work.

45. Richard R. Nelson and Sydney G. Winter, *An Evolutionary Theory of Economic Change* (Cambridge, MA: Harvard University Press, 1982).

46. Nathan Rosenberg, *Perspectives on Technology* (Cambridge and New York: Cambridge University Press, 1976), 63.

47. Devendra Sahal, *Patterns of Technological Innovation* (Reading, MA: Addison-Wesley Publishing, 1981), 19, emphasis in original.

48. The aggregation problem in the economics discipline was known as the Cambridge Capital Controversy that erupted during the 1960s and early 1970s. The controversial debates were basically about the methodological validity of the neoclassical school led by Paul Samuelson of Cambridge, Massachusetts, and the Keynesian school led by Joan Robinson of Cambridge, England, in describing how economic growth and development actually take place and how economists can theorize these phenomena.

49. Alan A. Walters, "Production and Cost Functions: An Econometric Survey," *Econometrica* 31 (1963): 1–66, 11.

50. Anwar Shaikh, "Laws of Production and Laws of Algebra: The Humbug Production Function," *Review of Economics and Statistics* 56 (1974): 115–120, and Herbert A. Simon, "On Parsimonious Explanations of Production Relations," *Scandinavian Journal of Economics* 89 (1979): 459–474. Only Shaikh's critique will be presented here.

51. For an extremely insightful critique of the aggregate production function presented in a historical context, see John McCombie, "Rhetoric, Paradigms, and the Relevance of the Aggregate Production Function," paper presented at the *Tenth Malvern Political Economy Conference,* August 1996. McCombie argues that the success of the Solow model to be the reigning paradigm of neoclassical growth accounting and its ability to withstand powerful critiques like Shaikh's can be traced to the superior rhetorical strategy adopted by Solow and his followers. Despite being a regressive research program, the Solow growth

accounting model survives because of the conservative nature of reigning paradigms and the power of the backers of the reigning paradigm in the economics discipline. For an interesting critique of the aggregate production function analysis of technological change, see Felipe, "Factor Productivity."

52. Lawrence J. Lau, "Technical Progress, Capital Formation and Growth of Productivity," in *Competitiveness in International Food Markets,* Maury E. Bredahl, Philip C. Abbott, and Michael R. Reed, eds. (Boulder, CO: Westview Press, 1994), 145–167; J. L. Kim and Lawrence J. Lau, "The Sources of Growth of the East Asian Newly Industrialized Countries," *Journal of the Japanese and International Economies* 8 (1994): 235–271; and Paul R. Krugman, "The Myth of Asia's Miracle," *Foreign Affairs* 76, no. 6 (1994): 62–78.

53. Felipe, *Factor Productivity.*

54. Paul Romer, "Idea Gaps and Object Gaps in Economic Development," *Journal of Monetary Economics* 32 (1993): 543–573, and "Endogenous Technological Change," *Journal of Political Economy* 98 (1990): S71-S102; and Robert E. Lucas, "On the Mechanisms of Economic Development," *Journal of Monetary Economics* 22, no. 1 (1988): 3–42. However, proper credit for "endogenizing" technological change by parsing the technological change index (residual) of Solow by allocating it into various changes in the economy should go to Solow's contemporaries Edward Denison, *The Sources of Economic Growth in the U.S. and the Alternatives Before Us* (New York: Committee for Economic Development, 1962), and John W. Kendrick *Productivity Trends in the United States* (Princeton, NJ: Princeton University Press, 1961).

55. Romer, "Idea Gaps." "Mixing" things in the right proportion can develop new materials. Technological innovation can be equated to the knowledge embedded in the heuristic of finding the right proportion. A good example would be the high temperature superconducting material (which is basically a ceramic material) by mixing the component materials in innumerable combinations until the "right" combination is found.

56. Jacob Schmookler, *Inventions and Economic Growth* (Cambridge, MA: Harvard University Press, 1966).

57. Rosenberg, *Perspectives on Technology.*

58. Zvi Griliches, "Hybrid Corn and the Economics of Innovation: A Critique," in *The Economics of Technological Change: Selected Readings,* Nathan Rosenberg, ed. (Harmondsworth, UK: Penguin Books, 1971), 212–228.

59. This criticism is valid for Schmookler's approach as well. For a detailed criticism of approaches that compartmentalize different components of technological change and isolate one factor and give credit only to that factor alone as responsible for technological change, see John F. Hanieski, "The Airplane as an Economic Variable: Aspects of Technological Change in Aeronautics, 1903–1955," *Technology and Culture* 14, no. 4 (1973): 535–552.

60. Unlike neoclassical economists, institutional economists believe that institutions are endogenous economic variables. Institutionalists use modern choice theory to analyze institutional change within the framework of "methodological individualism." For more details of the institutionalists' empirical and theoretical approaches, see Richard Grabowski "The Theory of Induced Institutional Innovation," *World Development* 16, no. 3 (1988): 385–394.

61. These analysts' views on induced innovation and technological change can be found in Vernon W. Ruttan, "Induced Institutional Change," in *Induced Innovation: Technology, Institutions, and Development,* Hans Binswanger and Vernon Ruttan, eds. (Baltimore:

Johns Hopkins, 1978), 327–357, and "Technical and Institutional Transfer in Agricultural Development," *Research Policy* 4 (1975): 350–378; Hans Binswanger, "Induced Technical Change: Evolution of a Thought," in *Induced Innovation,* Binswanger and Ruttan, eds., 13–43; and Vernon Ruttan and Yujiro Hayami, "Technology Transfer and Agriculture Development," *Technology and Culture* 14, no. 2 (1973): 119–151.

62. Binswanger and Ruttan, "Introduction," in *Induced Innovation,* 3.

63. Binswanger, "Technical Change," 19. Theoretically, this is very close to the notion of technology as knowledge and information, a concept much in vogue among historians of technology who reject the internalist paradigm the historians of science use to write the history of technology.

64. Binswanger, "Technical Change," 14.

65. Ruttan and Hayami, "Technology Transfer," and Ruttan, "Institutional Transfer."

66. Ruttan and Hayami, "Technology Transfer," 125.

67. Ruttan and Hayami, "Technology Transfer," 125.

68. Vernon W. Ruttan, "The International Agricultural Research Institute as a Source of Agricultural Development," *Agricultural Administration* 5, no. 5 (1978): 293–308, further identifies three separate submodels of institutional transfer of technology within the capacity transfer scheme. The temporary migration of experts to the recipient country to impart advice to local technical experts is called the "counterpart" model. The establishment of universities and institutes to train local experts to transfer know-how with the help of foreign benefactors is called the "university contract" model. The third submodel is called the "international institute" scheme whereby international research organizations transfer know-how and help to adapt it to fit the transferee's particular situations. Obviously, these observations seem to fit the case of technology transfer in the agricultural sector, more so than sectors like consumer goods and heavy industries.

69. The important works of these analysts where path dependence is discussed are: Nathan Rosenberg, *Exploring the Black Box: Technology, Economics, and History* (Cambridge: Cambridge University Press, 1994), *Inside the Black Box, Perspectives on Technology,* and "The Direction of Technological Change: Mechanisms and Focusing Devices," *Economic Development and Cultural Change* 18, no. 1 (1969): 1–24; and Paul A. David, *Technical Choice, Innovation, and Economic Growth* (New York: Cambridge University Press, 1975), and "Clio and the Economics of QWERTY," *American Economic Review, AEA Papers and Proceedings: Economic History* 75, no. 2 (1985): 332–337.

70. Rosenberg, *Perspectives on Technology,* 28–29.

71. Rosenberg, *Exploring the Black Box,* 10.

72. Rosenberg, *Exploring the Black Box,* 16.

73. David, "Economics of QWERTY."

74. The DSK system was patented in 1932 by August Dvorak and W. L. Dealey. The DSK system held most of world's records on speed typing. The Apple IIC PC had a built in switch to instantly convert its keyboard from QWERTY to virtual DSK. For details, see David, "Economics of QWERTY."

75. According to David, "Economics of QWERTY," the story of QWERTY begins with Christopher Sholes, a Milwaukee, Wisconsin, printer who was the fifty-second man to "invent" the typewriter. Sholes patented his "Type Writer" in October 1867 with help from his friends and collaborators, Carlos Glidden and Samuel Soule. Commercial success did not come to Sholes because of several problems with his design, such as the nonvisibility of the typed prints because the printing point was placed underneath the paper carriage.

Also, the tendency of the typebar to clash and jam if struck in rapid succession was a serious problem. Working for several years to perfect his machine, through trial-and-error, Sholes came up with the idea for a typewriter, which had more or less the same appearance as the modern-day four-row QWERTY system. In 1873, Shole's financial backer James Dunsmore succeeded in entrusting with E. Remington & Sons the manufacturing rights of the Sholes-Glidden typewriter. It appears that the only reason for the QWERTY layout to be adopted was to enable the salesmen to easily type the words TYPE WRITER to impress the customers. In late 1880s, Remington bought out the rights of the Sholes-Glidden patents, and the Sholes-Remington typewriter with the "universal" QWERTY layout was born. Subsequent efforts to change the layout to facilitate speed typing never succeeded to capture any market share.

76. David, *Technical Choice,* 334.

77. David, *Technical Choice,* 335.

78. Shaikh, "Laws of Production," Simon, "Parsimonious Explanations," and Felipe, *Factor Productivity.*

79. McCombie, "Rhetoric, Paradigms."

80. Planck quoted in Thomas Kuhn, *The Structure of Scientific Revolutions* (Chicago: University of Chicago Press, 1970), 151.

5

Neo-Schumpeterian
Evolutionary Models

TECHNOLOGY AND THE EVOLUTIONARY PARADIGM

After Karl Marx, Joseph Schumpeter stands out among those historically informed economics scholars who provided considerable insights on technological change as a key factor behind economic development and social change. Jon Elster calls Schumpeter "probably the most influential single writer on technical change."[1] Ironically, Schumpeter is dismissed as a mere business-cycle theorist by mainstream economists as one who had little to say about how economic growth actually takes place.[2] However, mainstream economists are slowly catching up with Schumpeter's insights to formulate new theories of economic growth by placing primary emphasis on the role of technological change.[3] Though Schumpeter gave unusual weight to the innovative process in the overall technological trajectory, his demystification of invention without justification has to be reevaluated (for a detailed analysis of these issues, see the next section). Nonetheless, his identification of technological change as the key to economic change and evolution is significant. Though Schumpeter postulated that technological innovation is a discontinuous and disruptive process, he saw it as unfolding according to an evolutionary norm.

The concept of evolution is not used, as it is understood in its biological sense of a slow, self-generating random process. However, it is used here essentially for its heuristic potential for locating the selection process.[4] As Joel Mokyr correctly points out, any analogy between Darwinian evolutionary process and technological change can be misleading. According to Mokyr, "[T]he evolutionary method in technological change is not so much an *analogy* with biology as another application of a Darwinian logic that transcends the world of living beings."[5] While the environment forms the selection background in biological evolution, the particular political economic context (broadly conceived as the sociocultural context and narrowly conceived as the market mechanism) acts as the selection background for technology. Despite the above caveats, however, many analysts do see a close parallel between the evolution of biological entities and technological artifacts, as will be seen in this chapter. John Smith, for example, claims that the best way to think about technological change is to view it as akin to a biological evolutionary process.[6] Technological change evolves as a result of the selection of technical novelty into a proper socio-economic-cultural niche. George Basalla warns that since we do not ascribe "progress" to biological evolution, technological evolution as progress may be ascribed only "within very restricted technological, temporal, and cultural boundaries and according to a narrowly specified goal."[7] However, his argument that "the advancement of technology must be disengaged from social, economic, or cultural progress" seems to be unwarranted as will be shown later.

In this chapter, the focus will be on models and theories of technological change that exploit the Schumpeterian postulates ("creative destruction" and discontinuity) on this phenomenon within an evolutionary framework.[8] The evolutionary metaphor for explaining technological change has been garnering considerable epistemological support from the evolutionary epistemology of Karl Popper and Donald Campbell as many scholars of different persuasions who are interested in technological change began to recognize, in their deliberations, the epistemic aspect of this process.[9] In his most explicit account of the theory of evolutionary epistemology, Campbell did impute the mechanism of blind variation and selection to abstract concepts as well as to living organisms. However, what is being emphasized in this paper is to focus on the cognitive nature of knowledge production and its change. Because the emphasis (in technological change) is on the cognitive aspect of technologies, technological change is seen as a process of adaptation and change, mediated by the human agent within the context of her sociocultural milieu. In this chapter, we will analyze several models of technological change that use both the Schumpeterian schema of technological change as a discontinuous (but an adaptive) process and the heuristic of evolution (in various degrees of adaptation). Before these cases are looked at, a detailed analysis of Schumpeter's theoretical works on technological change is presented as a backdrop.

SCHUMPETER AND TECHNOLOGICAL CHANGE

Schumpeter theorized that the key factor that drives (capitalist) economic systems is technological innovation that is spearheaded by entrepreneurs in direct correlation to business cycles. Economic evolution or economic growth takes place as a result of the "creative destruction" of existing mechanisms of production and exchange in the marketplace caused by the proliferation of new innovations spearheaded by entrepreneurs. The recent resurgence of neo-Schumpeterian theories of technological change is an enduring sign of the historical significance of Joseph Schumpeter's theoretical works on the dynamics of capitalist economics and its change. Schumpeter's theories of innovation and his theories of the impact of innovations on economic evolution or economic development have made great contributions to the development of new theories and models of technological change. The simplicity and supposedly transeconomic nature of the explanation offered by Schumpeter seem to be the reasons for the interest in applying Schumpeter's theories of innovation and entrepreneurship. Although mainstream (neoclassical) economists do not take Schumpeter's theories seriously for economic policy analysis,[10] Schumpeterian theories of innovation have, however, established a lasting niche in the theories and models of technological change and progress. The discontinuity that he emphasizes in the economic development process, spearheaded by the innovator-entrepreneur, attracted technology analysts to build models of technological change and progress.

Schumpeter considered innovation as the most important factor responsible for economic change, and he de-emphasized the role that inventions play in economic and industrial development. Claiming that innovation is the most important technological activity, Schumpeter argues that:

> Technological change in the production of commodities already in use, the opening up of new markets or of new sources of supply, Taylorization of work, improved handling of material, the setting up of new business organizations such as department stores—in short, any "doing things differently" in the realm of economic life—all these are instances of what we shall refer to by the term Innovation. It should be noticed that that concept is not synonymous with "invention." Whatever the latter term may mean, it has but a distant relation to ours.[11]

Schumpeter further claims that the motivation for invention has no economically relevant effect at all. Schumpeter sees innovation and invention in stark historical contrast. He further claims that:

> In many important cases, invention and innovation are the result of conscious efforts to cope with a problem independently presented by an economic situation or certain features of it, such as for example, shortage of timber in England in the sixteenth, seventeenth, and eighteenth centuries. Some *innovation* is so conditioned, whereas the corresponding *invention* occurred independently of any practical need. This is necessarily so whenever innovation makes uses of an invention or a discovery due to a happy accident, but also in other cases. It might be thought that innovation can be anything else but an effort to cope with a given situation.[12]

Unlike invention, according to Schumpeter, innovation is a "distinct internal factor of change" of the capitalist economic system. The "economic leadership" that carries forward the innovative functions is vested in the entrepreneurs.

Schumpeter defines innovation "rigorously" by means of a production function, and thus he sees "innovation as the setting up of a new production function."[13] The outward or inward movement of the isoquant will be the indication of the setting up of a new production function. However, Schumpeter insists that this shift should not be construed as a "change of techniques." The setting up of new plants, the generating of credits, and facilitating the emergence of entrepreneurs are the means by which innovation is operationalized. The emergence of new methods of production, according to Schumpeter, characterizes the act of "creative destruction."[14] During this process, the innovating entrepreneurs set up new methods of production to produce old goods at lower cost.[15] The emergence of factors like these destabilize the circular flow of the economic system, culminating in a new equilibrium, as more goods and services are produced and added to the system.[16]

In terms of their conceptualization of technological change represented by the production functions, Schumpeter and the neoclassical economists did not differ much. Schumpeter's version of innovation was basically the setting up of a production function.[17] In many respects, Robert Solow (see chapter four) developed his now famous growth accounting model of technological change by following the lead provided by Schumpeter. Thus Vernon Ruttan argues that the "current interest in technological change and growth of total productivity is focused on the same problem which Schumpeter treated under the heading of innovation."[18] Paul Schweitzer, one of the early analysts of Schumpeter's theory of technological change, correctly cautions that the technological change implied by the shift in the production function is "far too restrictive" because it excludes "changes in productive techniques other than those which result in shifts in the production function."[19]

As pointed out by Schweitzer, "change in productive techniques" also shifts the production isoquant. Thus, it is probable that the change in "productive technique" may be caused by invention as well, and hence, a shift of the production functions. The only difference is that the invention causes the shift in a more roundabout way. This is the case only if the invention is an intermediate good, such as a new machine or a new method of production. If the new invention is embodied in a final consumption technological good, then the shift is straightforward. Thus, Schumpeter's insistence that only innovation causes the shift of the production function may be incorrect. The only meaningful explanation for Schumpeter's belief that the shift of the isoquants due to a "change in technique" cannot form new production function is that the techniques are normally caused by new inventions.[20]

Following the Schumpeterian theory that innovations cluster, Chris De Bresson argues that the best way to induce technological "accumulation" is to breed and nurture a "network of innovating firms and industries."[21] Based on the Schumpeterian stricture, De Bresson argues that it is useful to distinguish between invention and innovation without offering why such distinction would enhance the conceptualization of technological change. De Bresson's model would have gained a lot more explanatory power had he included inventive activity along with innovative activity in the locus of the "technological trajectory" of the firms that foster "technological accumulation."[22] Using the United States biotechnology as a case study, Martin Kenney also claims that the Schumpeterian distinction between invention and innovation would be useful to explain the growth and development of the industry.[23]

Invention, innovation, transfer, and diffusion are different stages in the process of technological change.[24] Ruttan argues that invention, innovation, and technological change must be seen as a logical sequence: "That is, invention in some manner is antecedent to innovation, and innovation in turn antecedent to technological change."[25] New technologies may evolve and old technologies may change due to a variety of factors. Richard Nelson, while presenting a detailed survey of the literature on the "economics of invention," stresses that "invention is strongly motivated by perceived profit opportunities."[26] He further adds that the state of scientific, and presumably technological, knowledge significantly affects cost and demand factors, which in turn affect the "profitability of invention."[27] Nelson privileges invention in the same way that Schumpeter privileges innovation. Ironically, commenting on Schumpeter's distinction between invention and innovation, Nelson and Winter argue that "in the current institutional environment with much of innovation coming from internal R & D, the old Schumpeterian distinction is much less useful than it used to be."[28]

Jacob Schmookler argues that inventive activity is essentially an endogenous economic process, unlike Schumpeter's stricture that it is an exogenous factor.[29] Using inventive activities in numerous industrial fields, Schmookler shows that invention is a very important economic activity.[30] Nathan Rosenberg complains that the uncritical acceptance of Schumpeter's model of innovation as the major model by everyone interested in technology and economics resulted in a lack of consideration of activities other than major innovations that promoted productivity increase and economic growth. Rosenberg claims that the reason Schumpeter distinguishes invention from innovation lies in his concept of the discontinuous nature of innovative activity. The "clustering of innovations was at the heart of his business cycle theory."[31] Rosenberg wonders why Schumpeter had not examined invention as a "continuing activity whose nature, timing, and special problems are relevant to the subsequent Schumpeterian stages of innovation and imitation."[32] Rosenberg rightfully complains that by "creating artificial disjunction between innovative activity and other activities with which it is not only linked, but which in fact constitute major parts of the historical process of innovation itself," we have created major "intellectual barriers" to understanding the nature of technological change.[33]

Invention and innovation are both important technological activities. There are strong economic reasons to carry out inventions. Ruttan appears to be correct when he argues that to get out of the restrictive definitions of invention and innovation, we should abandon our attempts to provide analytically different definitions of these activities.[34] He further adds that we consider invention as "an institutionally defined subset of technical innovations" so that this conceptualization can accommodate the emergence of new technologies according to A. P. Usher's inventive schema of "cumulative synthesis."[35] Schumpeter's concept of innovation, once modified to include invention in the technological trajectory, may be a useful tool for studying the dynamics of technological change. The hypothesis that innovations catalyze economic development and that it follows an evolutionary trajectory are Schumpeter's greatest contributions to studying technological change. It is precisely these astute theoretical observations of Schumpeter that motivated subsequent analysts of technology to build interesting and important models of technological change. Thus, we should jettison his idiosyncratic notion that invention is outside the locus of technological development, and instead should concentrate on other important findings of Schumpeter for constructing models of technological change.

NEO-SCHUMPETERIAN AND EVOLUTIONARY MODELS

Institutional economists were the pioneers in using the evolutionary model to explain technological change. In this respect, Karl Marx, Joseph Schumpeter, and Thorstein Veblen should be considered the forerunners of this approach. Unlike the neoclassical economists, institutionalists see social and cultural factors as powerful institutional agents that spur on economic change. Attempting to formulate a theory of technology, Thomas DeGregori argues that technological change is essentially an evolutionary process.[36] Following the logic of a quasi-Darwinian evolutionary theory, DeGregori explains technological change as akin to the evolution of new species through the mechanism of natural selection. DeGregori sees technological change as a coevolutionary process along with human evolution.[37] In many respects it is analogous to the evolution of languages, as "the dynamics of tool using and of open-ended language are a function of the evolutionary process from which human beings emerged."[38] Technological evolution occurred as a dynamic feedback response to the biological evolutionary process. "In a very real sense," argues DeGregori, "technology evolves as the adaptation of the environment to the organism."[39] DeGregori's conceptualization of technology and technological change is close to the notional concepts of human evolution and tool development developed by paleoanthropologists.

DeGregori based his theory on field research conducted in developing countries as part of a technology transfer study mission. He argues that technological adaptation and borrowing between cultures has been going on since antiquity. For him, the opposition to the transfer of modern (western) technology to the Third World that is being shown by some development theorists is ahistorical and antiprogressive. DeGregori draws empirical support for his evolutionary model of technological change, mainly from the preindustrial period, specifically from the domain of agriculture. DeGregori claims that what is basic to technology is "ideas," a useful metaphor with which one can learn something about the processes of technology transfer and diffusion. However, finding an isomorphism between human evolutionary processes and technological evolution beyond an aphoristic level comes dangerously close to unwittingly endorsing sociobiology. Early tool development and usage may have coevolutionary features, but finding the same in modern societies is too simplistic and ahistorical. However, one claim that saves DeGregori from being branded a naive evolutionist is his attributing a purpose to technological evolution as opposed to the randomness in human evolution. The purpose of technological evolution is to improve human living conditions by "thoughtful, intelligent action directed toward problem solving."[40]

Another institutionalist who considers technological change analogous to biological evolution is Milton Lower.[41] Lower analyzes technological change from the Veblenian perspective of technology by treating technology as an aspect of culture and human knowledge. Lower considers technological change to be an aspect of cultural change and posits a major transcultural component to technology. This transcultural aspect of technology makes it possible to transfer and adapt it to other cultures according to their particular environment. Both Lower and De-Gregori reject the neoclassical economics position of technological change as being a narrow selection process involving only market forces. However, both assume technology transfers across time and space from one culture to another as an evolutionary process. They seem to be oblivious to larger historical forces, such as colonial and imperial conquests, the imposition of foreign technology on subjugated peoples, and the particular dynamics of the diffusion of imposed technology in the subjugated lands.[42]

Predominantly, it was the institutionalist economists who resurrected the insights of Schumpeter for analyzing and modelling technological change. As anticipated, the theoretical frame they adopted was evolutionary theory. On the one hand some took technological change as akin to biological evolution (examples: DeGregori and Lower see above) and on the other some took it as a heuristic device (Nelson and Winter, see below) to organize the existing empirical evidence on the process of innovation and technological change. Nelson correctly points out that the neoclassicists' way of conceptualizing technological advance as a maximizing process fails to address the point, as does the attempt to treat it as similar to a biological process.[43] Nelson and Winter take their basic theoretical starting point of technological change from a cultural-evolutionary perspective.[44] Any attempt to model technological change formally, as attempted in neoclassical economics, will end up as utter failure, because, at the core, it is an uncertain and contingent process. Because of the uncertainty and contingency, the production function model has only limited analytical scope, and one has to reject the ancillary hypotheses of profit maximization, perfect rationality, and utility maximization. Nelson rejects the maximizing postulate of neoclassical economics and adopts the satisficing postulate of Herbert Simon as the fundamental principle of his stochastic model.[45]

The recognition of "uncertainty" involving R & D forms the basic starting point for the Nelson-Winter model. Looking at the processes of innovation[46] in different industrial sectors, such as agriculture, industrial machinery, aviation, and so on, Nelson and Winter claim that, for "any hope of integrating the disparate pieces of knowledge about the innovation process, a theory of innovation must incorporate explicitly the stochastic nature of innovation, and must have considerable room for organizational complexity and diversity."[47] It is possible to locate these specific "heuristic search process" in particular industries or sectors, which they claim can be called an R & D strategy. Under certain conditions and constraints, facing the particular organization in question, this "strategy can be then represented in terms of a conditional probability distribution of innovations (or innovation characteristics)."[48] The heuristic, however, does not open up a readily usable way of operating this selection mechanism, unlike an algorithm where there is clearly a laid out procedure to follow, in order to get to the solution.[49]

Nelson and Winter claim that, in following this heuristic from the demand side, it may be possible to trace or locate a certain "natural trajectory" or some sort of an "internal logic" of innovation.[50] Conversely, they call natural trajectory, a "technological regime," which is similar in meaning to the "metaproduction function" of Hayami and Ruttan.[51] Metaproduction function is meant as a "frontier of achievable capabilities" that is characterized by a specific problem-solving culture. Nelson and Winter claim that their conceptualization of technological regime is "more cognitive, relating to technician's beliefs about what is feasible or at least worth attempting."[52] The example they posit for a technological regime is the Douglas DC-3 aircraft developed in the 1930s, whose basic design and overall features acted as the basis for future technological change in aeronautics and aircraft frame design. The term "technological regime" appears to be similar to a technological paradigm in the strict Kuhnian sense that Edward Constant uses to explicate the turbojet revolution in aeronautics.[53] A technological regime not only defines the boundaries of the area of investigation and problem-solving possibilities, but also defines the "design trajectories."[54] Certain natural trajectories could be common to many technologies such as economies of scale and accessibility to mechanization and automation.

The operationalization of identifiable R & D strategies, from which organizations have to choose a small subset of feasible routines, would inevitably lead to particular outcomes based on chances. Therefore, Nelson and Winter see technological change as a purposive, and yet intrinsically stochastic process taking place in a conceptual framework provided by a "selection environment" that is circumscribed by a technological regime. Markets, politics, regulatory demands, and diverse social institutions could act as the selection environment. In addition to expanding on the market selection environment, they do not shed enough light on the other factors and actors mentioned in their model. Martin Fransman argues that Nelson's and Winter's model of technological change falls short of providing a complete account of this complex phenomenon, and may be "seen as being in a 'preparadigmatic' stage of its development."[55] Fransman further claims that "while central aspects of the technical change process have been identified, these together with other economic, sociopolitical, and cultural aspects have not yet been rigorously welded into an acceptable theory of the determinants of technical change."[56] This criticism is unfair because it was Nelson and Winter who first came out with a clear heuristic to account for technological change by specifying factors other than purely economic ones.

Following the publication of Nelson and Winter's model, there came a proliferation of neo-Schumpeterian evolutionary models of technological change. Among these, a most prominent and promising model was that of Giovanni Dosi, who was at the time of publication of his famous paper associated with the Science Policy Research Unit (SPRU) at the University of Sussex.[57] Dosi legitimates the need for his new model by rightfully rejecting the "demand-pull" model popularized by neoclassical economists and "technology-push" model popularized by theorists of autonomous technology and technological determinism.[58] He constructs his model first by suggesting a two-part definition for technology, calling it as a piece of knowledge that has (a) a practical side characterized by embodied concrete devices and problems, and (b) a theoretical part characterized by know-how, methods, and procedures.[59] Secondly, he defines a "technological paradigm" as similar in methodological style to a Kuhnian scientific paradigm. A "technological paradigm" is a "model" and a "pattern of solution of *selected* technological problems based on *selected* principles derived from natural sciences and on *selected* material technologies."[60]

The parallel term for Kuhn's normal science in Dosi's model is "technological trajectory," which he defines as the "pattern of 'normal' problem-solving activity (i.e., of 'progress') on the ground of a technological paradigm."[61] Dosi also claims that in identifying a technological paradigm to chart the technological trajectory, one must specify a "positive heuristic" and a "negative heuristic." A positive heuristic will specify what direction to follow and the negative heuristic will specify what direction not to follow while carrying out technological tasks. What Dosi is trying to show is that technological paradigms have a powerful "exclusion effect." That is, engineers and R & D personnel often tend to be "blind" to certain possibilities of technological solutions, given generic technological tasks. For example, such generic technological tasks as in transportation, or in production of certain chemical compounds with specific properties, or in switching and amplifying electrical signals, the technological paradigms that emerged to solve the particular technological tasks in the above three examples were internal combustion engines, petrochemical processes, and semiconductors, respectively.[62]

While specifying the selection criteria, *ex ante,* for a particular technological paradigm, the evolutionary part of the model is highlighted among several notional possibilities. Some of the obvious selection criteria are marketability and profitability. Some less obvious, but potentially powerful, selection criteria are cost- and labor-saving possibilities and social and industrial conflicts.[63] Particular social conflicts that occurred during the design stages of certain machines were found to be profound determinants of new technologies.[64] It is possible to sort out, from each technological task or problem, the respective positive and negative heuristics. Dosi cautions that despite the emergence of a clear technological paradigm and its trajectory, powerful institutional factors can still influence the rate of change and growth of technological knowledge. Dosi also claims that the establishment of technological trajectories corroborates the presence of Kondratief's long waves (business cycles) that are in turn caused by the clustering of innovations as predicted by Schumpeter.[65]

Although Dosi cautioned that the epistemological analogy between scientific and technological change is meant to be "impressionistic," his adherence to the notion of paradigms is more than impressionistic.[66] The paradigm concept would run into problems when explaining the cumulative nature of technological change. Also, the assumption of competing technological paradigms existing in the same industry is rather tenuous, given the fact that firms and other economic agents motivated by profit or market share tend to use the same technology if patent restrictions are not operative. Henk Van den Belt and Arie Rip contend that they "seriously doubt the historical adequacy of the assumption of the plurality of competing technological paradigms" in the Dosian scheme of technological change.[67]

Leaving aside the excessive economic assumptions of Nelson and Winter's and Dosi's models, Van den Belt and Rip develop a "synthetic model" called the "Nelson-Winter-Dosi" model by extending and elaborating on the "sociological content" of their models. Van den Belt and Rip accurately preface their synthesis by cautioning that technological paradigms and technological trajectories conjure up images of technological determinism. They contend that the common denominator of evolutionary thinking between the Nelson-Winter-Dosi models and the social constructivist models that they follow is sufficient "ground for comparing and evaluating both approaches."[68] They argue that the best conceptual tool to analyze technological change is the "clustering of heuristics around an exemplary [technological] achievement."[69] Therefore, a technological achievement like the Douglas DC-3 aircraft can be considered as an "exemplar" in the Kuhnian sense. Van den Belt and Rip's grand synthesis of the Nelson-Winter-Dosi model can be put as follows:

> The appearance of an exemplar is a necessary but not sufficient condition for "normal" technological development along a trajectory to occur. In addition, there have to be expectations about the success of continuing work within this cluster of heuristics-expectations that must be embedded in the subculture of the technical practitioners and others involved in the development. . . . The combination of exemplar and cultural matrix forms a technological paradigm, and the further articulation of such a paradigm, partly influenced by the selection environment, leads to a technological trajectory.[70]

The focus should be on the nexus, the social institutions that interact between the selection environment and the trajectory. Van den Belt and Rip use the empirical case of the patent system in the synthetic dye industry as the nexus to elaborate their model. In the highly competitive synthetic dyestuff industry, looking for the *Muttersubstanz* or the skeleton of the dyestuff molecule became the heuristic for the innovation game for the firms and inventors. The patent system, as the social nexus or cultural matrix, is still too limited in explanatory power as the mediating force between the selection environment and the technological trajectory. The concept of heuristics still needs further articulation and elaboration, as Van den Belt and Rip's treatment of it is at times ambiguous and tenuous. Their attempt does have potential, provided Van den Belt and Rip can provide a new model of their own instead of using the skeleton of the Nelson-Winter-Dosi model of technological change—their *Muttersubstanz*.

Another neo-Schumpeterian evolutionary model of technological change that came out after the Nelson-Winter model was that of Devendra Sahal.[71] Sahal, an industrial engineer by training, claims that the two existing conceptions of technology, the production function (neoclassical economics) and the Pythagorean (by which it appears he means historical) are inadequate.[72] Sahal presents many of the problems of the production function model of technological change that were discussed in the literature (see chapter four). The Pythagorean concept of technology simply means counting relevant technological factors and events such as the number of patents issued, the number of inventions, and the number of engineering and technology graduates, and other measures of rating the pace of technological change in a nation. Sahal argues correctly that the production function concept is similar to the Pythagorean concept in that it also explains very little about the actual process of technological change.

Discounting these two concepts as inadequate, Sahal claims that the best alternative is a systems concept of technology.[73] Sahal offers four advantages of the systems view over the other two views. The first one is that the systems view offers a clear functional measure of technology that can be clearly defined and measured. For example, the thermal efficiency of an electric power plant can be defined as the ratio between energy output and input defined on a uniform measure of heat energy. The thrust-to-weight ratio of a jet engine can similarly be measured. The second one is the functional measures of technology that have more practical usage in actual innovative processes than economic measures such as prices. R & D personnel can decide what parameter of the technology is required in order to accommodate the particular need of the firm or the consumer. Thirdly, functional measures make it possible to account for both major and minor innovations by making it possible to assign appropriate composite measures of performance characteristics. For example, the fuel consumption efficiency measure of an automobile can be appropriately weighted to account for major innovations like pneumatic tyres, improved fuel mixing, and minor innovations like more durable pistons, valves, and body contouring. Fourthly, functional measures make it easy to detect changes in both the product and process characteristics of technology.

Having outlined the superiority of the systems concept of technology over the neoclassical production function and the supposedly historically oriented Pythagorean concepts, Sahal develops his systems-based model of technological progress and change. Sahal downplays seemingly revolutionary advances in technology and instead focuses on the continuous and cumulative nature of technological change. What is important for him is to follow the minute details of the technological development process in order to detect the formation of a system, which he expresses as follows:

> As regards the process of technological change, very often there emerges a pattern of machine design as an outcome of prolonged development effort. The pattern in turn continues to influence the character of subsequent technological advances long after its conception. Thus innovations generally depend on bit-by-bit modifications of a design that remains unchanged in its essential aspects over extended periods of time. This basic design is in the nature of a guidepost charting the course of innovative activity.[74]

The way to analyze technological change is to look for the "guidepost" and "innovation avenue" in the particular technological example in point. Sahal claims that it is possible to detect one or two designs standing out above others. The subsequent innovations are always based on these "invariant" designs.

Sahal uses the farm tractor as an example to substantiate his model. In the history of farm tractors, it is possible to identify two predominant designs, the Fordson and the Farmall. Sahal claims that all the subsequent innovations in farm tractors were based on these skeletal designs from the 1920s. From the aeronautics industry one can pick out the Douglas DC-3 aircraft as the invariant design on which all subsequent airplane innovations evolved. However, Edward Constant would object seriously to this analysis of the technological change in the aircraft industry, as we saw in chapter two. Sahal, on the contrary, claims that "there has been a large element of continuity in technological advances throughout the history of aircraft," despite the introduction of the jet engine.[75] Irwin Feller claims that Sahal's approach aims at integrating the macro level dimensions of the neoclassical production model and the micro level dimensions postulated by the evolutionary metaphor at the firm level.[76] Sahal employed two methods to "test" his model of technological change. The first method was to use econometric regression analysis, and the second method was to formulate a "lawlike relationship" between technology and economic agents. Feller argues that Sahal's econometric testing suffers from his uneven data on different studies of technological innovations, and that his "search for lawlike relationships is disquieting and most often unconvincing."[77]

Paolo Saviotti offered a more sophisticated evolutionary-systems model in the neo-Schumpeterian genre.[78] Saviotti claims that though economists consider technological change as the cause for economic growth and productivity increases only engineers and historians seem to understand the intrinsic properties of this process. In his systems representation of technology, the technical issues represent the internal structure, and the external or environmental characteristic is represented by the service modes of the technological artifact. Technological change, according to Saviotti, is the "process of adaptation of technologies to their external environment." The adaptation takes place with the internal variables kept relatively constant, similar to the concepts of self-regulation and homeostasis in complex electronics and biological systems, respectively. Saviotti claims that from this perspective, technological change models represented by "technological regimes," "natural trajectories," "technological paradigms," and "technological guideposts" can be interpreted as examples of "homeostasis of complex technological systems."

Saviotti does not clearly elucidate what he means by "technical factors" when he refers to the internal characteristics of a technology. In his model, he claims that the change takes place with the internal factors kept "relatively" constant. However, then what exactly happens when technological change occurs? Actual case studies of technological change clearly show that internal factors change considerably, and particularly the knowledge content of the technology and those of the community of technology practitioners. It seems that he extends the evolutionary metaphor literally to the case of technological change as a slow and steady but minor adaptation of the technological system to the external environment. The external environment concept is also rather nebulous in this formulation.

CONCLUSION

Schumpeter's pioneering attempts on explaining what causes business cycles in capitalist economic systems led him to theorize that such systems work because of the process of "creative destruction," which acts as their motive force. Creative destruction, again, is caused by the activities of innovating firms and entrepreneurs who seek new ways of doing old things or do new things. He argued that it is technological innovations that cause economic change, and the clustering of innovations indicated the empirical presence of business cycles. The hypothesis that innovations spur on economic change and that it follows an evolutionary trajectory is Schumpeter's greatest contribution to the study of technological change. However, Schumpeter's claim that invention is not an important variable in economic dynamics, not part of the technological nexus, needs re-examination. Building models of technological change by following Schumpeter's assertion of this invention-innovation dichotomy will potentially lead to a cul-de-sac. Schumpeter's hypotheses should be used for their general principles and their heuristic guidance for further studies and research on technological change.

It was the synthesis of the evolutionary epistemology with Schumpeterian notion of technological innovation that became an attractive theoretical option for building various models and theories of technological change. Technological change is a learning and an adaptive process. Models can be built, either formally or informally, by focusing on the artifact. By giving agency to the artifact or by focusing on the human agents and the community of technological practitioners, the agency question becomes less of a problem. Most of the models discussed above followed the former stricture. Basalla claimed that the primary unit of analysis of technological change must be the artifact. The discovery of blurring the boundary between animate and inanimate actors is a possibility that the meta-evolutionary technological model-builders missed. This radical constructivist position was the natural trajectory and guiding posts of heterogeneous engineer-sociologists and actor network builders.

As Elster astutely pointed out, both neoclassical models and evolutionary models tend to focus the level of technological activity at the firm level.[79] This tendency often deprives these theories of lessons offered by the larger historical synthesis offered by Marx and Schumpeter.[80] However, historically oriented analysts like Basalla and DeGregori do not restrict themselves to the firm or micro level. They invoke cultural and larger socioeconomic factors in their theories and models. Attributing isomorphism between biological evolution and technological evolution is untenable beyond a metaphorical transference of the concept of evolution. Primarily this is because evolutionary theory in biology offers a functional explanation, while in technology it has to be circumscribed by an intentional-cum-functional explanation where the randomness of "selection" becomes a meaningless concept. In order to remove any ambiguity that technological change is not isomorphous to biological evolution and also to accentuate the public nature of technological knowledge, Richard Nelson, therefore, argues that technological change should be conceptualized as a "cultural evolutionary" process.[81]

ENDNOTES

1. Jon Elster, *Explaining Technical Change: A Case Study in the Philosophy of Science* (Cambridge: Cambridge University Press, and Oslo: Universitetsforlaget, 1983), 138.

2. Paul Samuelson, "Schumpeter as a Teacher and Economic Theorist," *Review of Economics and Statistics* 33 (1951): 98–103, 98, Schumpter's most illustrious student, for example, writes that "Schumpeter will unquestionably be labeled by future historians of thought as a business cycle theorist who placed primary stress on the role of the innovator." Samuelson's tribute to his teacher was an insult in many ways, to categorize Schumpeter as a mere business cycle theorist and not as an influential economics theorist and thinker. However, Schumpeter's legacy seemed to have outlived his student's uncharitable appraisal of his work, as we will see in this chapter.

3. See the highly influential papers of Paul Romer, "Idea Gaps and Object Gaps in Economic Development," *Journal of Monetary Economics* 32 (1993): 543–573, and "Endogenous Technological Change," *Journal of Political Economy* 98 (1991): S71-S102, in which neoclassical economists seem to rediscover the role of technological change in spurring economic growth and development.

4. The far-fetched idea that technology is autonomous and technological change is a self-generating process is seriously mooted by some technological determinists. See, Jacques Ellul, *The Technological Society* (New York: Knopf, 1964). The evolutionary epistemology attributed to technological change has nothing in common with the theory of technological determinism.

5. Joel Mokyr, "Evolution and Technological Change: A New Metaphor for Economic History," in *Technological Change: Methods and Themes in the History of Technology,* Robert Fox, ed. (Amsterdam: Harwood Academic Publishers, 1996), 63–83, 64, emphasis in original.

6. John K. Smith, "Thinking about Technological Change: Linear and Evolutionary Models," in *Learning and Technological Change,* Ross Thomson, ed. (New York: St. Martin's Press, 1993), 65–78.

7. George Basalla, *The Evolution of Technology* (Cambridge, UK: Cambridge University Press, 1988), 216.

8. Theoretical insights from evolutionary models of technological change will be used in chapters six and seven while explaining the technological change in the Green Revolution in Indian agriculture.

9. Karl Popper, *Conjectures and Refutations: The Growth of Scientific Knowledge* (New York: Basic Books, 1963); and Donald T. Campbell, "Evolutionary Epistemology," in *The Philosophy of Karl Popper,* Paul A. Schlipp, ed. (La Salle, IL: Open Court, 1974), 413–463.

10. Schumpeter worked within the classical and the neoclassical economics tradition, but mainstream economists of all traditions (except those in the evolutionary/institutionalist tradition) did not take Schumpeter's works seriously. For more on these issues, see Robert Wolfson, "The Economic Dynamics of Schumpeter," *Economic Development and Cultural Change* 7 (1958/1959): 31–54.

11. Joseph Schumpeter, *Business Cycles,* 2 volumes (New York: McGraw-Hill Press, 1939), 84.

12. Schumpeter, *Business Cycles,* 85, note 1, emphasis in original.

13. Schumpeter, *Business Cycles,* 87.

14. Paul Strassmann, "Creative Destruction and Partial Obsolescence in American Economic Development, *Journal of Economic History* 14, no. 3 (1959): 335–349, argues that Schumpeter's description of "creative destruction" in which dominant new production methods completely overwhelm old methods of production may be inaccurate. He claims that old methods often coexist in an industry that is undergoing technological change. A new technology completely replaces an old technology only if the rate of obsolescence in the industry occurs with "unforeseen rapidity" (336).

15. Joseph Schumpeter, *The Theory of Economic Development* (Cambridge, MA: Harvard University Press, 1934).

16. In a nutshell, according to Wolfson, "Economic Dynamics," 45, Schumpeter's model "rests on the innovation process and the innovator, on the credit mechanism, and on the drive for profit maximization." In Carolyn Solo's words, "Schumpeter regards innovation as the truly dynamic element in the economy, the source of credit, interest, and profit as well as business fluctuations." See Carolyn S. Solo, "Innovation in the Capitalist Process: A Critique of the Schumpeterian Theory," *Quarterly Journal of Economics* 65, no 3 (1951): 417–428, 427.

17. In Schumpeter's version the input factors are land and labor, and in the neoclassical version these factors of production are capital and labor.

18. Vernon W. Ruttan, "Usher and Schumpeter on Invention, Innovation and Technological Change," in *The Economics of Technological Change: Selected Readings,* Nathan Rosenberg, ed. (Harmondsworth, UK: Penguin Books, 1971), 73–85, 76.

19. Paul R. Schweitzer, "Usher and Schumpeter on Invention, Innovation and Technological Change: Comment," *Quarterly Journal of Economics* 75, no. 1 (1961): 152–156, 153.

20. A more reasonable position might have been that of the neoclassical economists, which does not differentiate between invention and innovation at the production stage. Instead, this perspective looks at the increase in productivity due to technical change as a

result of the forward shift of the isoquant. This, however, does not mean that the neoclassical model offers a complete account of the process of technological change, as argued earlier in chapter four.

21. Chris de Bresson, "Breeding Innovation Clusters: A Source of Dynamic Development," *World Development* 17, no. 1 (1989): 1–16, 2.

22. J. A. Bombardier, the entrepreneur-innovator whom De Bresson identifies as an exemplar to support his model, can be identified as an inventor-entrepreneur as well. It is only a matter of convention as to when and how to identify Bombardier's automobile products as a result of either inventive or innovative activities. He has more in common with Thomas Edison, whom Thomas Hughes, *The Networks of Power: Electrification in Western Society, 1880–1930* (Baltimore: Johns Hopkins, 1983), considers as the greatest inventor-entrepreneur of all time.

23. Martin Kenney, "Schumpeterian Innovation and Entrepreneurs in Capitalism: A Case Study of U.S. Biotechnology Industry," *Research Policy* 15 (1986): 21–31.

24. See more details of this conceptual point in Nathan Rosenberg, *Perspectives on Technology* (Cambridge and New York: Cambridge University Press, 1976); Thomas Hughes, "The Development Phase of Technological Change," *Technology and Culture* 17, no. 3 (1976): 423–431; and Ruttan, "Usher and Schumpeter."

25. Ruttan, "Usher and Schumpeter," 73.

26. Richard R. Nelson, "The Economics of Invention: A Survey of the Literature," *Journal of Business* 32, no. 2 (1959): 101–127.

27. Nelson, "Economics of Invention," 101.

28. Richard R. Nelson and Sydney G. Winter, "In Search of a Useful Theory of Innovation," *Research Policy* 6 (1977): 36–76, 61.

29. Jacob Schmookler, *Inventions and Economic Growth* (Cambridge, MA: Harvard University Press, 1966).

30. Nathan Rosenberg, *Perspectives,* commenting on Schmookler's work, argues that the long-held tradition of treating invention as an exogenous activity outside the economic sphere has been reversed by the path-breaking work of Schmookler, despite his neglect of the supply responsiveness of inventive activities.

31. Rosenberg, *Perspectives,* 67.

32. Rosenberg, *Perspectives,* 67.

33. Rosenberg, *Perspectives,* 77.

34. Ruttan, "Usher and Schumpeter," 83.

35. Ruttan, "Usher and Schumpeter," 83.

36. Thomas R. DeGregori, *A Theory of Technology: Continuity and Change in Human Development* (Ames: Iowa State University, 1985).

37. It needs to be mentioned here that DeGregori, *Theory of Technology,* does not use the term "coevolution" in his book.

38. DeGregori, *Theory of Technology,* 11.

39. DeGregori, *Theory of Technology,* 14.

40. DeGregori, *Theory of Technology,* 24.

41. Milton Lower, "The Concept of Technology within the Institutionalist Perspective," in *Evolutionary Economics,* Volume I, Marc R. Tool, ed. (Armonk, NY: M. E. Sharpe, 1988), 197–226.

42. For a detailed exposition of technology transfer through imperialist campaigns, see Michel Adas, *Machines as the Measure of Men: Science, Technology, and the Ideologies*

of Western Dominance (Ithaca, NY: Cornell University Press, 1989); Claude A. Alvares, *Decolonizing History: Technology and Culture in India, China and the West 1492 to the Present Day* (New York: Apex Press, and Goa, India: Other India Press, 1991); and Daniel H. Headrick, *The Tools of Empire: Technology and the European Imperialism in the Nineteenth Century* (New York: Oxford University Press, 1981).

43. Richard R. Nelson, "Technical Change as Cultural Evolution," in *Learning and Technological Change,* Ross Thomson, ed. (New York: St. Martin's Press, 1993), 9–23.

44. See their interesting book, Richard R. Nelson and Sydney G. Winter, *An Evolutionary Theory of Economic Change* (Cambridge, MA: Harvard University Press, 1982).

45. Elster, *Technical Change.*

46. Nelson and Winter, "Useful Theory," 36, use the term "innovation as a portmanteau to cover the wide range of variegated processes by which man's technologies evolve over time."

47. Nelson and Winter, "Useful Theory," 48.

48. Nelson and Winter, "Useful Theory," 52.

49. A good strategy is to follow both the demand side and the supply side factors in picking out inventions and innovations.

50. Nelson and Winter argue that "natural trajectory" may be similar to "technological imperatives" or "technological bottlenecks" that Nathan Rosenberg uses as the guiding principles of technological change.

51. Hayami and Ruttan's metaproduction function relates output (Q) to inputs capital (K), land (N), labor (L), human capital (H), and research and development (R), such that:
$$Q = f(K, L, N, H, R).$$
For details of the metaproduction function, see Yujiro Hayami and Vernon W. Ruttan, *Agricultural Development: An International Perspective* (Baltimore: Johns Hopkins, 1971). It may be recalled that this equation has some similarity to the endogenous neoclassical production function model.

52. Nelson and Winter, "Useful Theory," 57. Attributing the task of technological development work only to "technicians" reveals a lack of knowledge on the part of Nelson and Winter of how technological development actually takes place. Technological development is a multidimensional process involving more agents than technicians.

53. Edward W. Constant, *The Origins of the Turbojet Revolution* (Baltimore: Johns Hopkins, 1980).

54. Nelson and Winter, "Useful Theory."

55. Martin Fransman, "Conceptualising Technical Change in the Third World in the 1980s: An Interpretive Survey," *Journal of Development Studies* 21, no. 4 (1985): 572–652.

56. Fransman, "Technical Change," 610.

57. Giovanni Dosi, "Technological Paradigms and Technological Trajectories," *Research Policy* 11, no. 3 (1982): 147–162.

58. In its purely neoclassical version, "demand-pull" model basically means that in order to meet the recognized needs of the consumers, producers would move to satisfy those needs through appropriate technological efforts. Essentially, the market acts as the signaling device on an a priori basis to activate inventive and other technological activities. Relative prices and quantities would be the regulatory mechanism, among other market indicators. "Technology-push" is just opposite of the "demand-pull" model of technological change. Here, technological change is purely a deterministic process, dictated by the internal needs and demands of the particular technological systems.

59. Dosi, "Technological Paradigms," 151–2.

60. Dosi, "Technological Paradigms," 152. Dosi equates technological paradigm to a "research programme," a concept propounded by Imre Lakatos, *The Methodology of Scientific Research Programmes* (Cambridge: Cambridge University Press, 1978), in the philosophy of science.

61. Dosi, "Technological Paradigms," 152.

62. For detailed analyses of each of these problems and the search process that led to the particular solutions, see Dosi, "Technological Paradigms."

63. Dosi, "Technological Paradigms," 155.

64. For industrial conflicts as determinants of new technological developments, see David Noble, *Forces of Production: A Social History of Automation* (New York: Knopf, 1984), from the machine tool industry during the Cold War period in the United States; and Tine Bruland, "Industrial Conflict as a Source of Technical Innovation: Three Cases," *Economy and Society* 11 (1982): 91–121, about industrial manufacturing and textile industries in Britain and the United States during the eighteenth and nineteenth centuries.

65. Dosi, "Technological Paradigms," 160.

66. Despite his caveats, Dosi's exposition of a "downward" sequence of "science-technology-production" assumption is ahistorical and unhelpful to understand the evolution of technological paradigms. Dosi argues that along this downward sequence, economic forces (along with institutional and social factors) operate as the selection device. See Dosi, "Technological Paradigms," 153.

67. Henk Van den Belt and Arie Rip, "The Nelson-Winter-Dosi Model and Synthetic Dye Chemistry," in *The Social Construction of Technological Systems: New Directions in the Sociology and History of Technology,* Wiebe E. Bijker, Thomas P. Hughes, and Trevor Pinch, eds. (Cambridge, MA: MIT Press, 1987), 135–158.

68. Van den Belt and Rip, "Nelson-Winter-Dosi," 136.

69. Van den Belt and Rip, "Nelson-Winter-Dosi," 140.

70. Van den Belt and Rip, "Nelson-Winter-Dosi," 140.

71. Sahal's developed his model in the following works. See Devendra Sahal, "Technological Guideposts and Innovation Avenues," *Research Policy* 14 (1985): 61–82, *Patterns of Technological Innovation* (Reading, MA: Addison-Wesley Publishing, 1981), and "Alternative Conceptions of Technology," *Research Policy* 10 (1981): 2–24.

72. Sahal, "Alternative Conceptions."

73. According to Sahal, *Technological Innovation,* 25–27, the systems view of technology was first conceived by systems analysts concerned with the management of R & D in industrialized countries, and by development economists interested in finding "appropriate technology" for meeting the needs of Third World countries. Apparently, the factor in common between these two disparate groups is their interest in functional parameters such as efficiency, thrust-to-weight ratio, and so on of modern technology. This claim is highly inaccurate, as historians of technology had used a systems concept of technology for a considerable length of time, as we saw in chapter two.

74. Sahal, *Technological Innovations,* 33.

75. Sahal, *Technological Innovations,* 35–6.

76. Irwin Feller, "Review of *Patterns of Technological Innovation* by Sahal, D.," *Science* (2 July 1982): 47.

77. Feller, "Review of *Patterns,*" 47.

78. Paolo P. Saviotti, "Systems Theory and Technological Change," *Futures* 18, no. 6 (1983): 773–786.

79. Elster, *Technical Change.*

80. Elster, *Technical Change,* 10.

81. Nelson, "Cultural Evolution."

6

Technological Change as Problem Solving

INTRODUCTION

It has been argued effectively by anthropologists, sociologists, economists, and others that the reason we show considerable interest in technological change as a process mediated by our social, cultural, and cognitive attributes is because of its practical implications. To be specific, the reason is the promise to improve the material living conditions of humans. Every society undergoes technological change with varying degrees of intensity. The many technologies that we take for granted have come from elsewhere. The transfer and diffusion of technologies from one nation/society to another, either voluntarily or though coercive means (such as imperialism), has been going on since antiquity.[1] Technological change, or lack of it, was responsible for bringing about fundamental shifts in the economic fortunes of peoples and nations as Smith, Marx, Schumpeter, and others had shown. We attempted to understand this phenomenon through various models and theories presented in previous chapters. We will show in this chapter how technological change can be understood as a problem-solving activity undertaken by agents (economic and social) with a clear policy framework in mind. We will use the Green Revolution in Indian agriculture as the empirical basis for this purpose. Before presenting a detailed case study of the Green Revolution, we will look at a brief account of the historical context within which this agricultural modernization program took place.

MODERNIZATION AND TECHNOLOGICAL CHANGE

The modernization project, a legacy of the Enlightenment, was carried out through industrialization and economic development for more than two hundred years. The Industrial Revolution was the materialization process of the modernization project. This has completely transformed the social relations of production and consumption of most people in the industrialized nations. The process by which the feudal relations of production changed to a modern factory-based capitalist mode of production was, in reality, a change in the knowledge base of production intended for solving the particular needs and problems of agents (loosely defined), and is referred to as technological change.

The process of modernization in Europe began with the onset of a new productive relationship engendered by the new industrial and military technologies.[2] The concomitant transformation of agriculture and the rapid expansion of small-scale craft production into large-scale commercial agriculture and urban-based factory manufacturing created a radical shift in the way technological knowledge production had been organized and effected. This transformation in the means and modes of production along with the Enlightenment Project initiated the process of modernization. With the onset of modernization in Europe, a modern world system evolved that emerged out of the ruins of the medieval feudal social order. According to Immanuel Wallerstein, as a result of this transformation a European world economy came into being that was based upon the capitalist mode of production.[3] What made this new world system unique in world history was that the ideological force that impelled it was the economic domination and exploitation of other peoples, their lands, and their resources.[4] The ideology that sustained imperialism and the concomitant power associated with it enabled a few European nations to conquer much of the globe. This power was derived largely from their technological superiority.[5]

While fulminating against the inequities of colonialism and imperialism, Marx, an ardent believer in the modernization project, hoped that the transfer and diffusion of such modern technologies as telegraph, railways, electricity, and mechanized production methods would dramatically alter the feudal social relations that existed in the colonies. However, Marx's expectations were not realized because the technological change and the ensuing industrialization of the colonies were only peripheral, and most often unarticulated. Rather, these new technologies were only transferred for further military conquests, pacification, and normalization of the colonies.[6] The technological might of the colonizers was intended to cement their dominance over the colonies and perpetuate the notion of the superiority of the West and its technology. This situation began to change as a result of the massive decolonization process that began during the post–World War II period. The modernization project with the infusion of modern technology through different economic development models began in earnest during the 1950s and onwards in most of the former colonies, also collectively called the Third World.

It appears that for the decolonized Third World nations, the two "natural" models of economic development and industrialization were the Western "free market capitalism" and the Eastern "planned state capitalism." The common denominator of these two models was modern technology, the rapid infusion of which was expected to materialize through its transfer from the respective camps. While most

Third World countries opted for the former model, several others went with the latter. India, for example, opted for the former, but retained facets of the latter by incorporating features of planning into its development program. The modernization project also considered Western aid (or Eastern aid depending upon the case) in capital and technology as vital for achieving development. The basic assumption of modernization theory is convergence, an important ontological premise of the whole project that first appeared in Europe during the Enlightenment. That is, the world is on a particular path of progress and change engendered by the ideals of the Enlightenment; the West arrived there first, and the Third World would reach there eventually through a catching-up process.[7]

It is axiomatic in modernization theory that Third World ("traditional") societies can be transformed through a concerted project of economic development, which can be achieved by changing the means of production (technology) and by transforming and remolding archaic social structures that resist technological change in these societies. Changes in the means of production would entail a change in the relations of production, either to a free market or to a planned (statist) system of economic organization. It must be added, however, that the latter model is in complete retreat and many nations that followed it have reverted to the former after the collapse of the Soviet Union. Modernization can thus be achieved by adopting the "right" policies by the government. By formulating and implementing the "right" package of policies, the state and other agents of economic power can induce technological change, where technological change can be equated to a problem-solving activity. This minimalist, though profoundly effective, model can be a useful heuristic to understand technological change. The objective of this chapter is to demonstrate this hypothesis, by using the Green Revolution in Indian agriculture as the empirical focus. First, a historical reconstruction of the Green Revolution is provided.

TECHNOLOGICAL CHANGE IN INDIAN AGRICULTURE: THE GREEN REVOLUTION

The term Green Revolution refers to the changes in agriculture technology and mode of practice of agriculture experienced by some Third World countries during the late 1960s and 1970s. As a result, these countries experienced considerable increase in the production of basic staple cereals like wheat and rice.[8] It is an instance of a relatively successful technology transfer, in terms of increases in per capita cereal production. It presents a circumscribed way of looking at a narrowly defined national technology policy objective of increasing agricultural productivity.[9] Generally, the Green Revolution involved the use of seeds of high-yielding varieties (HYVs), primarily of wheat and rice, and the adoption of a "modern" package of improved agricultural practices involving chemical fertilizers, tractors, pesticides, controlled water, mechanical threshers, electric and diesel pumps, and so forth. These changes were instituted in place of the "traditional" agricultural practice involving the use of seeds whose genetic composition goes back thousands of years. "Traditional" technologies also include wooden plows, water-wheels, and bullock carts. In the traditional mode, animals and humans provided the energy required for all agricultural activities. Finally, traditional agriculture is dependent on the vagaries of monsoon rains.

Subsistence farming is often characterized by an "exclusion effect,"[10] that is, a tendency on the part of peasant farmers to resist modernization and rapid technological change, specifically, to resist radical innovations. This tendency to maintain the status quo, and consequently not to undertake innovations, culminated in depressed agricultural productivity, which in turn prompted the Indian government to formulate and implement a new agricultural policy to break out of the stasis in agriculture. The situation was compounded by the failure of monsoons that culminated in a near-famine situation in the 1960s.[11] In fact, the decline in agricultural production was caused by the newly independent Indian government, which placed the primary emphasis on rapid industrialization by setting up huge industrial plants that Nehru, India's first prime minister, called the "temples of modern India."[12] The agricultural sector was left to fend for itself and was expected to provide the surplus to sustain the emerging industrial sector. The decline in food production due to negligence and the lack of cooperation from nature (droughts) forced the government to import several million tons of subsidized food to avert massive famines. This frightening food situation forced the government, under heavy external pressure from aid donors and the World Bank, to introduce a new agricultural policy for increasing cereal production.

The agricultural policy was aimed at increasing land productivity by introducing a technological solution in the form of the Green Revolution package of technology. This was partly developed indigenously and was partly transferred from the West to India. There was no intention of introducing land reforms or changing agrarian relations. The policy objective was to target small-to-medium farmers who were encouraged to adopt the new agricultural practices. As there was no more land to be brought under the plow, increasing the productivity of the land using modern technology became the most viable means for providing food for nearly one-fourth of the world's population, who were dependent on only one-sixteenth of its land area.[13] The Green Revolution has made India self-sufficient in food grains *production* even though it has spread only to a quarter of its land area (in states such as Punjab and Haryana and parts of Uttar Pradesh, Karnataka, Tamil Nadu, and Andra Pradesh). The Green Revolution may be characterized as the new technological paradigm that replaced the old paradigm characterized by subsistence farming.

Succinctly put, the technological change in Indian agriculture may be seen as the transformation of the newly derived knowledge in agricultural technology, both Indian and foreign, into food. The Green Revolution is more than the development and diffusion of the HYVs of seeds, although there is a tendency on the part of many to equate the HYVs as synonymous with the Green Revolution. However, the HYVs form a key artifact in the Green Revolution technological network. HYVs are defined as early maturing semidwarf types that, under intensive agricultural practices (involving chemical fertilizers, irrigation, pumps, threshers, and so on), provide a significantly higher yield compared to the traditional types. The widespread belief that the HYVs originated in the West is incorrect.[14] Semidwarf rice varieties originated in China around A.D. 1000.[15] These caught the attention of Japanese farmers in the late 1800s because their productivity increased dramatically when chemical fertilizers were applied.[16] The plant improvement process began with selection in the late 1800s and the hybridization process in the 1900s. During the interwar period, as rice became a scarce commodity in Japan, the imperial government decided to transfer the new seed-fertilizer-irrigation-based technology to its colonies in Korea and Taiwan.[17] Semidwarf rice was highly productive in Korea, but was not successful in Taiwan because the warmer climate there was not conducive to the seeds that were developed for temperate Japan. The semidwarf varieties that evolved from the Japanese experience attracted the attention of other countries that wanted to use them as parents in new breeding programs.

The HYVs of rice suited to tropical conditions in Southeast and South Asia were developed in the 1960s at the International Rice Research Institute (IRRI) in the Philippines. The earlier varieties developed at the IRRI were based on genetic materials drawn from China, Taiwan, and Indonesia.[18] The most famous variety was known as IR-8 (Chandler 1973).[19] It was based heavily on experience in developing the Norin variety in Japan and Ponlai variety in Taiwan.[20] These HYV rice seeds were extensively and successfully introduced in several Third World countries in the 1960s. The most popular variety introduced in India, besides IR-8, was the Taichung Native 1, a related strain of IR-8.[21]

Semidwarf wheat originated in Japan in the 1800s, and the two most important varieties used for international breeding programs were Akakomugi and Daruma.[22] The Japanese crossed Daruma with several American varieties: the most productive variety that arose from these experiments was known as Norin 10. Other important varieties developed from Daruma were Seun 27 and Suweon 92. Norin 10 was introduced into the United States in 1946 and was crossed with several native varieties by the U.S. Department of Agriculture scientists.[23] In 1948, scientists in Washington State crossed Norin 10 with Brevor, a native variety. In 1954, the Norin-Brevor cross was taken to Mexico (to what is now the International Maize and Wheat Improvement Center, or CIMMYT known by its Spanish acronym), where Norman Borlaug and his colleagues developed several varieties of the HYVs of wheat seeds that were later transferred to India and other Third World countries in the mid-1960s.

A reconstruction of the history of the Green Revolution in India shows that four protagonists played crucial roles in its success.[24] They are the government of India, multilateral and bilateral donor agencies, international agricultural research institutions, as well as the farmers and peasants of India. The institutions under the government of India that planned and co-ordinated the transfer and diffusion of the new technology were the Ministry of Food and Agriculture and the Indian Council of Agricultural Research (ICAR),[25] along with the various agricultural research institutes and agricultural universities scattered all over India.

The multilateral and bilateral donor agencies were the Ford Foundation, the Rockefeller Foundation, the World Bank, and the United States Agency for International Development (USAID). USAID played a key role as liaison and financial supporter to United States land-grant universities in helping to set up agricultural universities in India.[26] The establishment of agricultural universities, patterned after the land-grant universities of the United States, is an important event in the history of the Green Revolution that helped to transfer "modern" agricultural knowledge from the United States to India. Three United States institutions played key roles in the development of modern agricultural research capacity in India. USAID helped with investments and logistical support to start up the land-grant type universities, the Rockefeller Foundation helped with the development of a national agricultural research system, and the Ford Foundation helped with farm extension work.[27] The international agricultural research institutions were the International Rice Research Institute (IRRI) and the International Maize and Wheat Improvement Center (CIMMYT).[28] Several more international agricultural research institutes were established after these, the most recent one being an international biotechnology research institute in Rome. In 1971, all the international agricultural research institutes were brought under the World Bank's control.[29]

The farmers and peasants of India who, by adopting and adapting the new agricultural technology to their particular situation, made the Green Revolution a possibility filled a key role in the Green Revolution saga. The development and spread of the Green Revolution involved different learning processes that will be discussed later. Although the technological change in Indian agriculture, in parts of the country, during the period from 1965 to 1975 is generally referred to as the Green Revolution, a reconstruction of the history of the technological change demands that we look at the developments in Indian agriculture beginning in 1952. We can observe three distinct stages in the process of the change.

The first stage (1952–65) was the development of a new and indigenous national agricultural research system. The year 1952 was chosen as the beginning of this period because it was in that year the first USAID-university contract was signed. Later, in the mid-1950s, the Indian government sought the assistance of the Ford Foundation, the Rockefeller Foundation, and USAID for establishing more agricultural universities and for developing a high-quality graduate school at the Indian Agricultural Research Institute in New Delhi.

The overhaul and reform of the agricultural bureaucracy in India marked the second stage (1962–67) in order to facilitate the transfer and diffusion of the HYVs to the Indian situation. A thorough revamping of the agricultural institutions and commodity committees established by the British colonial administration was an important facet of the newly developed research system. In 1962, Indian scientists successfully tested HYV Mexican wheat under Indian conditions; HYV rice was tested in 1964. HYVs were introduced for the first time nationally during the growing season of 1965–66.[30]

The third stage (1965–75) was marked by the change in agricultural practice as a result of the introduction of the HYVs. The farmers began to adopt the new technology extensively beginning in the 1965–66 growing season. Agricultural productivity increased steadily until 1975. After 1975, the increase in productivity began to taper off and levelled off on an S-shaped productivity curve. By this time, in parts of India, farmers had realized a considerable increase in their yield in comparison with the 1965 base. Agricultural productivity again began to show a considerable increase in the mid-1980s, which some analysts identify as the "Second Green Revolution."[31] Details of the technological change in relation to these stages highlighting the roles played by the protagonists mentioned earlier will be presented below.

In 1948, a year after independence, the government of India set up the University Education Commission to remold the educational system in India.[32] The commission recommended the establishing of "rural universities" similar to the land-grant universities of the United States. Subsequently, several high-powered committees consisting of Indian and American agricultural scientists and educators appointed by the Indian government recommended strengthening indigenous research efforts and college-level training as well as establishing agricultural universities in all the states of the Indian Union.[33]

The first land-grant type agricultural university was established in Pant Nagar, Uttar Pradesh, and is now known as Gobind Ballabh Pant University of Agriculture and Technology.[34] This university was modelled after the University of Illinois.[35] Other US land-grant universities (Kansas, Ohio, Missouri, Pennsylvania, and Tennessee) entered into partnership arrangements with the government of India to establish several other agricultural universities.[36] USAID provided the contract funds and represented the United States on behalf of the land-grant universities. According to Hadley Read, hundreds of researchers and agricultural experts from these universities went to India to help with the establishment of the universities and their research facilities.[37] Also, thousands of Indians trained in the United States returned to India to teach and conduct research.

The American system of a federally supported agricultural research system appealed to the Indian government. Having a federal structure, the Indians wanted to adapt the American system of first developing a federal agricultural research capability and then supporting state-level research projects. The last joint Indo-US Agricultural Research Review Team recommended some sweeping changes in the largely civilian-dominated bureaucratic agricultural research system. Because of pressure from the powerful Indian Administrative Service (IAS), the concerned ministers of agriculture were not able to implement the recommendations until C. Subramaniam took over as India's minister of agriculture in 1964. Subramaniam single-handedly overhauled the agricultural bureaucracy in India, giving the power and credibility to the agricultural scientists.[38]

Subramaniam took three critical steps. The first was to remove the permanent civil service officer, who was serving as the director-general of the Indian Council of Agricultural Research (ICAR), and to employ an eminent agricultural scientist, B. P. Pal, as the new director-general. His second step was to appoint a committee of agricultural scientists to review the latest joint Indo-US expert committee to recommend the necessary measures to reorganize the scientific research infrastructure. On the recommendations of the new team, Subramaniam incorporated all the independent research institutes and commodity committees under ICAR. The third step was to establish an Agricultural Research Service, similar to those existing in the United Kingdom and the United States, to provide guaranteed career paths to agricultural scientists.[39]

Convinced of the high-yielding capability of the new seeds, Subramaniam, along with his trusted agricultural experts, proceeded with the decision to distribute the HYV seeds to the peasants and farmers and also to show them the new agricultural methods. They distributed the seeds to farmers at highly subsidized rates for sowing during the 1965–66 growing season. Having reorganized the agricultural research system in India, Subramaniam and his team of agricultural scientists and bureaucrats decided to demonstrate the effectiveness of the new HYV seeds to the farmers. Massive public information campaigns about the new technological package using radio, press, and cinemas were organized by the government around the country in 1966. But it was not an easy task to convince the farmers to switch to the new technology because, having no precedent, they did not want to bear the risk of a disastrous harvest again after several droughts. The farmers were willing to take the risk only after they were convinced that the new HYV-based agricultural practice was more productive than their age-old one. Hence, in 1965 it was decided to launch a thousand demonstration programs all over the country.

According to this plan, a minimum of two hectares of each selected field was devoted to the new agricultural system. These parcels of land were entrusted to the extension officers and agricultural scientists. These government officials were given the task of demonstrating the effectiveness of the new technological practice as a model farm to the community. The farmers and peasants learned the techniques of using HYVs and acquired the knowledge to change their agricultural practices. The demonstration programs were successful, and the demand for the "miracle" seeds soared, resulting in the importation of 18,000 tons of HYV wheat seeds from Mexico, the largest seed transfer of its kind in human history.[40] The Green Revolution originated from this point. The yield of wheat doubled in 1970, four years after the introduction of the HYVs on a national scale.[41] Demonstration programs for the new rice varieties were conducted on a national basis, though not as widespread as for the wheat varieties. According to Norman Borlaug, India's decision to import and implement the new HYV-based technology set off a chain reaction, not only in India, but in Pakistan and elsewhere.[42]

The seeds were subsequently changed to accommodate the Indian environmental conditions and consumer preferences. The IR-8, the most popular rice variety, for example, was found to be chalky and sticky. Similarly, the Mexican wheat was of reddish hue; Indians preferred the amber and white wheat varieties. Indian agricultural biotechnologists were able to develop new seeds to satisfy the tastes and cultural preferences of the consumers, while at the same time retaining the genetic quality that guaranteed high cereal productivity. Modifications were also required of the varied soil and climate conditions in India, which differed from the conditions in Mexico for the wheat seeds and in the Philippines for the rice varieties. The agricultural experts took the feedback from the farmers and consumers seriously and were successful in the development of two Mexican wheat lines that "performed better in the field and the kitchen." Similarly, adaptive research on the rice seeds obtained from the Philippines yielded 221 varieties by 1983. In the 1983–84 growing season, 76 percent of the land under wheat and 54.1 percent of the land under rice were using HYVs.[43]

Technological change originating outside India was successfully transferred because of the urgent need for more food. The demand for the technological change in peasant agriculture was created by the failed development policies of the government. The model of economic development that the Indian planners pursued in the beginning was biased against agriculture. Agriculture was neglected until the rude wakening of the 1960s when the country was threatened by famine. The technological change in agriculture that ensued, from the introduction of a new package of agricultural practices, had been a direct result of the interaction and the formation of a successful network. This was comprised of the newly transferred HYVs, irrigation canals, research institutes, international organizations, universities, farmers, and extension agents. The effort to solve a major crisis in agricultural production, in turn, enabled the country to become self-sufficient in food grains, and to develop a highly successful agricultural research capability of its own.[44] The agricultural story is thus a "prime example of mutually reinforcing foreign and local research efforts."[45]

TECHNOLOGICAL CHANGE AS PROBLEM SOLVING

Although the exact nature of the trajectory and outcome of technological change cannot be predicted in advance, the temporal dynamics of this process shows one important feature: that technological change follows an evolutionary course and that it is possible to steer the course of development of technology in a "desirable" way by influencing the selection environment.[46] Two key theoretical insights about technological change that we have learned from the previous chapters are the cumulative and evolutionary nature of this process. By cumulative, it is meant that ideas, practices, theories, and laws from the past do pass on to the development of newer technologies.[47] The reason why this aspect tends to be neglected, or does not seem apparent, is due to the tacit nature of technological knowledge. Many technological practices and craft traditions are not committed to writing.[48] In many instances, they are transmitted through succeeding generations either in informal settings like shop floors and apprenticeship programs or in family craft practices. These sources, however, do not exhaust the whole of technological knowledge. With the integration of engineering, organized research, and development activities, along with journals and books on technological development and design practices, a large segment of modern technological knowledge production takes place in organized settings.

The concept of technological evolution is structurally and practically different from that of biological evolution. In the case of technological evolution, the "selection" process is characterized by instruction, understanding, "hands on" experience, and finally, cognitive change (for more on these lines of reasoning, see the next chapter). The term "evolution" is used here as an explanatory metaphor or metamodel. Technological evolution is not isomorphic or analogous to biological evolution. Technological change is construed as a selective-retention process that is adapted to a sequential process of variation and selection.[49] Within the milieu of sociocultural evolution, adaptive learning and perception lead to the accumulation and change of technological knowledge. Fundamentally, technological knowledge is gained by trial and error, learning by doing, learning by imitating, and learning from mistakes. The evolutionary concept is important in explaining technological change because it captures the temporal nature of this phenomenon.

The cumulative nature of technological change, on the other hand, implies that technological change is irreversible. By cumulative, it is not meant a uniformitarian theory of accretion of everything past from the history of that particular technological tradition. The major idea here is that the functional attributes and basic design principles and operational guidelines of a technological system, to a large extent, remain invariant. Unlike in the (natural) sciences, where paradigmatic changes may occur because of new experimental discoveries and revolutionary theoretical advances, in technology, the fundamental functional attributes remain more or less invariant. The end use of a wooden plow or a mechanical tiller is the same. The latter does the work faster using mechanical power, and the former does it more slowly by using human and animal power. The functional attributes of the technology remain the same in both cases.[50] This, however, does not preclude the interpretative flexibility of both artifacts in divergent ways because of the enrollment of new actors into the network. Technological change is fundamentally influenced and molded by its antecedents in a path-dependent way.[51] The idea of technological improvement as a means for efficient action remains the same in all societies. That is, more out of less. Improved efficiency, increased productivity, less cost, less human intervention for avoiding hazardous conditions, and so forth are achieved by improving on the existing technologies. Therefore, it is imperative that all models of technological change account for both the cumulative and evolutionary nature of this process.

From the above vantage point, technological change can be characterized as a problem-solving activity. Although it is a problem-solving activity conducted in need-based or need-induced circumstances, the solutions do not simply emerge of a concatenation of technologies postulated by such ahistorical theories as the production function model discussed earlier. Extant political, economic, social, and institutional factors, including government policies, organize the problem-solving activity, but by themselves do not provide the solutions. The technological problems vary from one situation to another and their solutions vary according to the degree of complexity of the technological system. Some of the technological problems are low efficiency, adverse environmental conditions, simple functional failures, imbalances between artifacts of different vintages, and inadequate organizational structures. These problems can be the direct or indirect result of climatic and geographic constraints, natural disasters, social and cultural demands for change, simple economic wants, military demands, varying resource positions, and many other contingent factors.

In the Green Revolution example we saw that all of the above, except military demands, were the focusing devices that influenced or molded the alteration and variation processes for the selection of the new agricultural technology. The selection environment was created by the combined efforts of the international donor agencies, the government of India and its various departments and offices, and the international and national research institutes. The transfer, diffusion, location specific adaptation, and the indigenous development of the new technological knowledge ultimately transformed the existing "traditional" knowledge system in areas where the Green Revolution made a lasting impact. The case clearly delineates an active selective-retention process at the three stages of the Green Revolution. Although not as explicitly apparent as during the first two stages, the third stage (1965–1975) vindicates this claim. During this period a distinct pattern of change in the practice of agriculture emerged as a result of the introduction of the HYVs and other new inputs.

The three stages in the history of the Green Revolution include understanding the new techniques and problems, learning new ways of doing things, and organizing a new research system. It is important to note also that the sociocultural background of the participants (actors) mediated the learning process described earlier. The technological change involved here follows a clear heuristic in which the development part and the diffusion part joined together and interactively brought about the change.[52] There evolved a technological algorithm of how new knowledge could be transformed into material outputs, in this case into food.

The technological algorithm evolved as a means for simplifying the complex knowledge of a new agricultural system into a simplified form that farmers and peasants could understand easily in order to produce the desired results (outputs). In simple terms, it can be equated to a decision-rule-making process that includes all the protagonists associated with the Green Revolution. The process of change was a conceptual-cum-cognitive change. The peasants learned the new agricultural practices and techniques by trial-and-error, by observing how the new HYVs were sowed and planted, and by learning from the experts sent by the government. For them the new agricultural technology became meaningful only after internalizing new ways of doing things. Technological change here involved not just the introduction of a few new artifacts, but a fundamental change in their knowledge related to agricultural technology and its practice itself. After they acquired the new technological knowledge, they in turn provided valuable feedback to the experts, who were able to accommodate their suggestions and complaints. Unless we understand how the change in knowledge actually takes place, technological change cannot be conceptualized fully. Without such a perspective, models of technological change become models of one technological artifact replacing another.

The role of the government and international bilateral and multilateral agencies was to create a proper selection environment for the technology users and developers, in this case, peasants-farmers, local extension workers, and research and development personnel. Unlike natural selection, where chance occurrences are the norm, in technological selection, persuasion and the perception of the need for the new knowledge decide the outcome. The possibilities for variation are limited and fixed, *ex ante,* to a large extent. As a result, the public agencies can guide the path of selection in a desired fashion in most instances where contingencies do not bedevil the policy objectives. Adverse and unexpected effects of the new technology might create the need for a new selection environment, which might change the direction of the technological trajectory.[53] Thus, to a large extent it is within the control of humans to determine how technological change is directed and molded.

It is apparent from the Green Revolution that, because the actors who make the decisions on matters related to technological selection were specific individuals and groups of people in society and because the creators of the selection environment were motivated by political and economic reasons, the technological change that occurred was both socially constructed and political-economic in nature. Governmental regulations, market forces, consumer preferences, and environmental factors weigh heavily during the course of technological development. It becomes clear that technological change in the present case was shaped by the active social, political, and economic agendas of the government and its allies.

While the political-economic and social interests set the stage for technological change and influence its trajectory, the actual process of learning and the cognitive processes required for that change were dictated by the cultural and cognitive attributes of the members of the particular social groups involved. Through premeditated and clearly planned actions they tried to change the behavior of nature to satisfy their needs. Although social, political, economic, and other factors provided the background for change, the internal dynamics of the actual change were determined by the functional attributes of the technology. These are increased possibilities of energy substitution, mechanization, the ability for reproduction of products and services based on the same ideas, and the inception and sustainability of action through methods of communication and control. Finally, it is important to recognize the decision-rules that the agents establish in order to decide between alternative courses of action.

The Green Revolution illustrates a classic social history of technological change. It is more than a chronicle of the development, transfer, and diffusion of artifacts and "techniques." Contingent and chance occurrences can be observed if we look at the historical development of the HYVs.[54] Significantly, the HYVs were only a component of the Green Revolution network. The case study demonstrates that technology entails a sound knowledge of the ways and means of doing things to solve a problem. The Green Revolution package of technology was a system of knowledge intended to circumvent the constraints of the land's low productivity. It cannot be separated from knowledge because what essentially took place after the technological change was a change from an old knowledge system to a new one designed to solve a problem that the old one failed to solve adequately.

The creation of this new technological system, which was largely developed abroad, was achieved by transferring a relatively "successful" form of knowledge to India. This included the development of a research system to adapt that knowledge to the specific needs and conditions of India, and its transmission to the users (mostly peasant-farmers), who ultimately learned the new knowledge through trial-and-error, learning by doing, learning by using, and other learning processes. These internal factors demonstrate that the technological system entailed ideas and thoughts at the one end and techniques and things at the other. Although these processes involved artifacts and techniques, they were only part of the technological system. It was the knowledge content of these artifacts and techniques that mattered. If the recipients failed to understand how these artifacts worked, how to reproduce them, and failed to adapt them to the recipient's specific circumstances, the transferred technology would have become extinct once the artifacts had worn out or were used up.

The peasant-farmers learned the new technology through different learning processes, including model demonstration programs, trial-and-error, formal and informal training programs, and training and visits from the extension agents, among other activities. The change from the old system to the new was not discontinuous. The changeover from old seeds to the HYVs was gradual. The peasant-farmers experimented with the HYVs by first devoting only a small fraction of their land to the new seeds. The peasant-farmers accepted the new seeds and new methods of farming only after they were convinced of the high productivity gains from the new agricultural system.

The driving force behind technological change is the limitation of the old technology and the constraint that is imposed on it by its surroundings. Increased productivity, improved efficiency, less cost, less human intervention, and other such technological improvement measures are normally achieved by improving on the existing technologies. The Green Revolution shows this facet of technological change. The knowledge change involved new and better ways of doing things. However, the fundamental nature of agricultural practice did not change. The changes were related to using new seeds and seed preparation, weeding and using chemical pesticides, watering, adding chemical fertilizers, and other such activities.

Droughts and other "natural" disasters, inadequate organizational structures, preindustrial technologies, social and cultural demands for change, simple economic wants, and the varying resource positions of regions and states were the problems that influenced the variation processes for the selection of the Green Revolution package of technology. The selection environment was created by the joint efforts of the government of India, various international donor agencies, and different national and international research institutes. We saw that the creation, transfer, and local adaptation of the new technology ultimately transformed the existing tradition-based knowledge system. The case clearly delineates an active selective-retention process at the three distinct stages of the Green Revolution.

The first stage (1952–1965) was the development of a national research system. The second stage (1962–1967) was the transfer and diffusion of the HYVs. The third stage (1965–1975) was the change in agricultural practice as a result of the introduction of the HYVs. These three stages in the Green Revolution include understanding the new techniques and problems, the organization of a new research system, and learning new ways of doing things by trial-and-error. The new selection environment or metamarket was largely constructed by the government and its allies during the third stage.

CONCLUSION

The Green Revolution corroborates the assumption that technological change can be understood as a problem-solving activity. The actions undertaken were a clear response to a specific problem that was recognized and identified, and for which a specific plan of action was implemented. As described earlier, the transfer, innovation, and diffusion of a new knowledge system solved the problems faced by the government and the selective group of citizens. It is easy to conceptualize the technological change that has taken place here by looking at it as a problem-solving activity. The political-economic and institutional factors organized the problem-solving activity. The technological change described as the Green Revolution demonstrates that it occurred as the outcome of a specific set of concrete actions and learning experiences of a wide array of actors. It elucidates the way in which problems were defined and solutions were sought through strategic planning activities of the government.

The inability to increase productivity beyond the subsistence level using the old technology was a direct correlate of its functional problems. The constraints of the old system provided the problems for which the new research and development activities were conducted. The solution was to come up with concrete measures to circumvent the constraints, low productivity due to technological stasis. It started with the seed. The old seeds were replaced by the hybrid-HYV varieties. Occasionally, chance occurrences of new strains were logically followed up by proper problem-solving techniques. The problems associated with using these varieties further necessitated new approaches to agricultural practices. Constant renewal of seeds, increased usage of fertilizers, water, and other inputs were mandated by the change of seeds. To reiterate, we see that the learning processes involved actions that necessitated substituting mechanical power for human and animal power, chemical fertilizers for manure and compost, HYVs for traditional seeds, irrigated water for rainwater, and so forth. The Green Revolution shows that human actions and learning experiences were responsible for the increase in cereal productivity rather than some *deus ex machina* from which these material benefits emanated.

The Green Revolution reveals the evolution of a technological algorithm of how new knowledge can be transformed into material output in the form of food. The algorithm was a means to simplify complex knowledge into a form of information that the peasant-farmers could understand. The peasant-farmers learned the new technology by trial-and-error, by observing how the new technology was prac-tised, and by learning from the extension workers. For them, the new agricultural technology became meaningful only after learning new ways of doing things. The claim that technological change is knowledge change is clearly manifested by the fact that their cognition of agricultural practice changed, about which more in the next chapter. This was not attributable to the introduction of a few new artifacts, but to a systemic change in the knowledge related to agriculture, which could be characterized as a problem-solving activity.

ENDNOTES

Some of the material in this chapter is drawn from my article, "Technological Change as a Problem-Solving Activity," *Technological Forecasting and Social Change* 40, no. 3 (1991): 235–247. Acknowledgment is duly accorded to Elsevier Science for the necessary permissions.

1. For an interesting discussion of this point, see Barry K. Gills and Andre G. Frank, "The Cumulation of Accumulation: Thesis and Research for 5,000 Years of World System History," *Dialectical Anthropology* 15 (1990): 14–42.

2. It is instructive to note here that before the onset of the Industrial Revolution in Eu-rope, several critical technologies and scientific ideas reached Europe before the sixteenth century from China, India, and Arabia. These included the magnetic compass, printing and paper making, the water mill, cast iron, iron-chain suspension bridge, piston bellows, met-allurgy, the loom, the lathe, gun powder, paper, much of mathematics, chemistry, and me-chanics. For extensive treatment of these issues, see Joseph Needham, *Science and Civi-lization in China,* Volumes 1–7 (Cambridge: Cambridge University Press, 1954–1988), and *The Grand Titration: Science and Society in East and West* (London: Allen & Unwin, 1969); Lewis Mumford, *Technics and Civilization* (New York: Harcourt Brace Jovanovich, 1963); A. Rupert Hall, "Epilogue: The Rise of the West," in *A History of Technology,* Vol-ume III, Charles J. Singer, E. J. Holmyard, A. Rupert Hall, and Trevor J. Williams, eds. (Oxford: Clarendon Press, 1957), 709–721; Charles Susskind *Understanding Technology*

(Baltimore: Johns Hopkins, 1973); Claude A. Alvares, *Decolonizing History: Technology and Culture in India, China and the West 1492 to the Present Day* (New York: Apex Press, and Goa, India: The Other India Press, 1991); Michael Adas, *Machines as the Measure of Men: Science, Technology, and Ideologies of Western Dominance* (Ithaca, NY: Cornell University Press, 1989); and William McNeill, "How the West Won," *New York Review of Books* XLV, no. 7 (23 April 1998): 37–39. These technologies and scientific ideas radically transformed economic production and social relations in western European societies, paving the way for the birth of industrial capitalism. The Europeans were pragmatic and astute, adopting and adapting these transferred technologies to their advantage. According to Rupert Hall, "Epilogue," 716, "Perhaps European civilization could not have progressed so rapidly had it not possessed a remarkable faculty for assimilation—from Islam, from China, and from India." These new technologies of production, transportation, and warfare, which evolved after the assimilation of the newly transferred technological knowledge from the East, enabled the Europeans to venture out of their imagined space to the real place and space of the present Third World.

3. Immanuel Wallerstein, *The Modern World System I: Capitalist Agriculture and the Origins of the European World-Economy* (San Diego, CA: Academic Press, 1974).

4. Adas, *Measure of Men;* and Wallerstein, *World System.*

5. Adas, *Measure of Men.*

6. Adas, *Measure of Men;* Alvares, *Decolonizing History;* and Daniel H. Headrick, *The Tools of Empire: Technology and European Imperialism in the Nineteenth Century* (New York: Oxford University Press, 1981).

7. This metamodel of modernization and the ensuing universalist narration is under tremendous attack from postcolonial and postmodern theories of development and change. It is beyond the scope of this chapter to become involved in this debate. However, for a synoptic view of this debate pertaining to technology, see Govindan Parayil, "Transcending Technological Pessimism: Reflections on an Alternative Technological Order," Working Papers in the Social Sciences, No. 43, Division of Social Science, Hong Kong University of Science and Technology, 28 December 1998.

8. Govindan Parayil, "The Green Revolution in India: A Case Study of Technological Change," *Technology and Culture* 33, no. 4 (1992): 737–756; and Hans P. Binswanger and Vernon W. Ruttan, eds., *Induced Innovation: Technology, Institutions and Development* (Baltimore: Johns Hopkins, 1978). The term "Green Revolution" itself was coined by William Gaud in a speech entitled "The Green Revolution: Accomplishments and Apprehensions" given at the meeting of the Society for International Development in 1968. For more details on this naming episode, see Dana G. Dalrymple, "The Adoption of High-Yielding Varieties in Developing Nations," *Agricultural History* 53, no. 4 (1979): 704–726; and Vernon W. Ruttan and Hans P. Binswanger, "Induced Innovation and the Green Revolution," in *Induced Innovation,* Binswanger and Ruttan, eds., 358–408.

9. It is beyond the scope of this chapter to go into a complete account of the social and economic impact of the Green Revolution in India, albeit the extreme seriousness of these effects. The Bhopal gas explosion, which has directly or indirectly killed over 15,000 people during the past 13 years, is definitely connected to the Green Revolution as the Union Carbide factory was set up to produce synthetic pesticides. As a result of the change in agricultural practice, hundreds of thousands of agricultural workers are killed and injured by agricultural machinery, chemical fertilizers, and pesticides. These "mini-Bhopals" occur

because the workers operate and handle these dangerous substances and tools without proper guidance and protective gears. For some details of these issues, see Govindan Parayil, "The 'Revealing' and 'Concealing' of Technology," *Southeast Asian Journal of Social Science* 26, no. 1 (1998): 17–28.

10. Giovanni Dosi, "Technological Paradigms and Technological Trajectories," *Research Policy* 11, no. 3 (1982): 147–162, defines "exclusion effect" in a different context involving technological paradigms as a tendency on the part of engineers and R & D personnel to be "blind" to other possibilities of technological innovation besides the one they "select" as the technological solution to a given technological problem.

11. The peasant-farmers' tendency to resist innovation beyond a reasonable measure is understandable given the risks to which they are already exposed such as the vagaries of the monsoon rains, flooding, droughts, and so on. Unless they are guaranteed of alternative income or food source, peasant-farmers would not undertake a new technological practice, as will be shown later.

12. According to Mohanlal L. Dantwala, "From Stagnation to Growth: Relative Roles of Technology, Economic Policy and Agrarian Institutions," in *Technical Change in Asian Agriculture,* Richard T. Shand, ed. (Canberra: Australian National University, 1973), 259–281, growth in agricultural production in India up to the year 1965 was disappointingly low, with famines still possible. Although the failure of agriculture to meet the needs of India from the time of independence in 1947 until 1965 reflected a neglect in favor of the industrial sector, according to Barrington Moore, *Social Origins of Dictatorship and Democracy* (Boston: Beacon Books, 1966), and George Blyn, *Agricultural Trends in India, 1891–1947: Output, Availability and Productivity* (Philadelphia: University of Pennsylvania Press, 1966), this failure should be contrasted to the deplorable agricultural situation in India during the British colonial period. According to the estimates of S. R. Sen, *Growth and Instability in Indian Agriculture,* Agriculture Situation in India XXI (New Delhi: Ministry of Food and Agriculture, 1967), the average annual food grain increase in India from 1901 until 1947/48 was a meager 0.3 percent. In fact, P. K. Mukherjee and Brian Lockwood, "High Yielding Varieties Programme in India: An Assessment," in *Technical Change in Asian Agriculture,* Richard T. Shand, ed., (Canberra: Australian National University Press, 1973), 51–79, show that during the 1930s the population growth rate outstripped the growth of food grain. Whatever increases there were in agricultural production came from increased acreage, which ceased early on as there was no more arable land to be brought under the plow. Famines were a frequent occurrence in British India. According to Charles Bettelheim, *India Independent* (New York: Monthly Review Press, 1968), the most recent large-scale Bengal famine during 1942–43 left nearly 3.5 million Indians dead. Although the production of basic staples declined during the colonial period, there was a resurgence in productivity of export-oriented cash crops such as cotton, tea, coffee, jute, rubber, and spices during the colonial period. For more details on the agriculture situation during the colonial period, see Christopher J. Barker, "Frogs and Farmers: The Green Revolution in India and Its Murky Past," in *Understanding Green Revolution: Agrarian Change and Development Planning in South Asia,* Tim P. Bayliss-Smith and Sudhir Wanmali, eds. (Cambridge: Cambridge University Press, 1984), 37–53; and Rakesh Mohan, D. Jha, and Robert Evenson, "The Indian Agricultural Research System," *Economic and Political Weekly* 8, no. 13 (1973): A21–A26.

13. The Green Revolution did make India self-sufficient in food production. However, millions of Indians are hungry every day because of their inability to buy enough food. No

famine had occurred in India recently on any large scale. Famines of small-scale nature are still prevalent in some poor states like Orissa and Bihar, where most of those affected are the forest dwellers and indigenous "tribal" populations.

14. Ruttan and Binswanger, "Green Revolution."

15. Dana G. Dalrymple, "The Development and Adoption of High-Yielding Varieties of Wheat and Rice in Developing Countries," *American Journal of Agricultural Economics* 67, no. 5 (1985): 1067–1073.

16. The major problem with "traditional" varieties of seeds was that beyond a certain point, productivity decreases below what it would have been if no chemical fertilizers had been applied. The plant absorbs the nutrients to increase the size of the straw, thus posing the danger of "lodging," which is a tendency to fall over because of top-heaviness. Semi-dwarf types, on the contrary, do not lodge. The increased fertilizer used is returned in the form of higher cereal yields. That is why new seeds are required for the new agricultural practice.

17. Yujiro Hayami, "Elements of Induced Innovation: A Historical Perspective for the Green Revolution," *Explorations in Economic History* 8 (1971): 445–472.

18. Ruttan and Binswanger, "Green Revolution."

19. The IR-8 was a short, stiff-strawed, and highly fertilizer-responsive variety. It yielded five to ten tons per hectare in India. However, the consumers disliked it because of its stickiness and chalky taste. Indian rice breeders later developed new varieties using the genetic material of the IR-8 that were palatable to Indian tastes, without losing the high-yielding potential of this new "miracle" seed variety.

20. Hayami, "Induced Innovation."

21. Robert F. Chandler, "The Scientific Basis for the Increased Yield Capacity of Rice and Wheat, and its Present and Potential Impact on Food Production in the Developing Countries," in *Food, Population, and Employment: The Impact of the Green Revolution,* Thomas T. Poleman and Donald K. Freebairn, eds. (New York: Praeger, 1973), 25–43.

22. Dana G. Dalrymple, "Changes in Wheat Varieties and Yields in the United States, 1919–1984," *Agricultural History* 64, no. 4 (1988): 20–36.

23. Dana G. Dalrymple, *Development of the High-Yielding Wheat Varieties in Developing Countries* (Washington, DC: United States Agency for International Development, 1986).

24. "Success" may be a highly contestable term here. Of course, it is meant from the point of view of the dominant actors in society such as the government, international donor agencies, and the medium-to-small-scale farmers. It cannot be called successful from the point of view of the marginal farmers who lost their meager land holdings due to debt or other reasons. It is beyond the scope of this chapter to go into these details.

25. The ICAR was established by the British colonial administration in 1929 under the name Imperial Council of Agricultural Research to conduct research on commercial crops for export. For details, see Mohan, et al., "Research System." After independence, the ICAR was renamed the Indian Council of Agricultural Research, ironically with the same acronym.

26. Hadley Read, *Partners with India: Building Agricultural Universities* (Urbana-Champaign: University of Illinois Press, 1974).

27. Read, *Partners with India,* 97–101; Lele and Goldsmith, "Agricultural Research," 313.

28. The IRRI, instituted in 1960, was the joint effort of the Ford Foundation and the Rockefeller Foundation. The Philippine government provided the land for the institute. The

CIMMYT, instituted in 1967, was also the joint effort of the Ford and Rockefeller Foundations, and the Mexican government provided the land. See for details, Vernon W. Ruttan, "The International Agricultural Research Institute as a Source of Agricultural Development," *Agricultural Administration* 5, no. 4 (1978): 293–308.

29. The World Bank controls these institutes by bringing them under the umbrella of the Consultative Group on International Agricultural Research (CGIAR).

30. Uma J. Lele and Arthur A. Goldsmith, "The Development of National Agricultural Research Capacity: India's Experience with the Rockefeller Foundation and Its Significance for Africa," *Economic Development and Cultural Change* 37, no. 2 (1989): 305–344; and C. Subramaniam, *The New Strategy in Indian Agriculture* (New Delhi: Vikas Publishing House, 1979).

31. See Raj Chengappa, "Agriculture: A Golden Revival," *India Today* (April 15, 1989): 78–83; and Surinder Sud, "State of Agriculture: India Poised for Take-Off," *Times of India,* 19 January 1989, 7(N).

32. Read, *Partners with India.*

33. Albert H. Moseman, *Building Agricultural Research Systems in the Developing Nations* (New York: The Agricultural Development Council, 1970).

34. C. Prasad, *Elements of the Structure of Terminology of Agricultural Education in India* (Paris: Unesco Press, 1981).

35. See Paul R. Brass, "Institutional Transfers of Technology: The Land-Grant Model and the Agricultural University at Pantnagar," in *Science, Politics, and the Agricultural Revolution,* Robert Anderson, et al., eds. (Boulder, CO: Westview Press, 1982), 103–163, and D. P Singh, "Agricultural Universities and Transfer of Technology in India: The Importance of Management," in *Science, Politics,* Anderson, et al., eds. 165–178, for a detailed account of the institutional transfer process, which culminated in the establishment of this land-grant university patterned after the University of Illinois.

36. Read, *Partners with India.* These universities are Punjab Agricultural University (Ohio State University), Haryana Agricultural University (Ohio State University), University of Udaipur (Ohio State University), Madhya Pradesh Agricultural University (University of Illinois), Orissa University of Agriculture and Technology (University of Missouri), Maharashtra Agricultural University (Pennsylvania State University), Andhra Pradesh Agricultural University (Kansas State University), and Mysore University of Agricultural Sciences (University of Tennessee).

37. Read, *Partners with India.*

38. Lele and Goldsmith, "Agricultural Research," 322.

39. Subramaniam, *New Strategy,* 21.

40. Subramaniam, *New Strategy,* 47–48.

41. David W. Hopper, "Distortions in Agricultural Development Resulting from Government Prohibitions," in *Distortions in Agricultural Incentives,* Theodore W. Schultz, ed. (Bloomington Indiana University Press, 1978), 69–78, 69. Mahmood, *The Myth of Population Control: Family, Caste, and Class in an Indian Village* (New York: Monthly Review Press, 1972), 61, in his field research in a Punjabi village, found that wheat yields often tripled for many farmers after the switch over to the "miracle" seeds. It should be mentioned that Mamdani's project was concerned about the effectiveness of top-down family planning programs that were implemented in the Punjabi villages during the time of the Green Revolution. His original research interest was not, however, the Green Revolution.

42. Norman Borlaug, *The "Green Revolution," Peace and Humanity* (Mexico City: CIMMYT Report and translation series no. 3, 1972).

43. Dana G. Dalrymple, *Development and Spread of High-Yielding Rice Varieties in Developing Countries* (Washington, DC: United States Agency for International Development, 1986), and Dalrymple, *Wheat Varieties.*

44. It must be pointed out that despite bumper crops, millions of Indians go hungry every day because of their poverty, which makes them unable to command enough food at the marketplace.

45. Lele and Goldsmith, "Agricultural Research," 328.

46. For theoretical details on selection environment, see chapter five. Johan W. Schot, "Constructive Technology Assessment and Technology Dynamics: The Case of Clean Technologies," *Science, Technology & Human Values* 17, no. 1 (1992): 35–56, presents an innovative approach to influencing and shaping the selection environment for the development of "cleaner" and "safer" technologies by using the methodology of "constructive technology assessment."

47. Edwin T. Layton, "Technology as Knowledge," *Technology and Culture* 15, no. 1 (1974): 31–41.

48. This facet of technology prompted Derek J. de Solla Price, "Is Technology Historically Independent of Science: A Case Study in Statistical Historiography," *Technology and Culture* 6, no. 4 (1965): 553–568, 561, to identify technology as "papyrophobic" and science as "papyrocentric," 49. The intellectual foundation for this evolutionary epistemology was first presented in Donald T. Campbell, "Evolutionary Epistemology," in *The Philosophy of Karl Popper,* Paul A. Schlipp, ed. (La Salle, IL: Open Court, 1974), 413–463, essentially expanding on the pioneering works of Popper on the evolutionary nature of knowledge creation and its change. Karl Popper, *Conjectures and Refutations: The Growth of Scientific Knowledge* (New York: Basic Books, 1963), delineates the essential epistemological problems associated with the acquisition and growth of knowledge, particularly of scientific knowledge.

50. Mostly, the operating procedures, materials for constructing the artifacts, and the energy sources for operating the artifacts are changed.

51. Nathan Rosenberg, *Exploring the Blackbox: Technology, Economics, and History* (Cambridge: Cambridge University Press, 1994).

52. Invention, innovation, and the development of the technologies are included in the "development" part, and the transfer and its local adaptation by the recipients are included in the "diffusion" part.

53. The selection environment, as Richard R. Nelson and Sidney G. Winter, "In Search of a Useful Theory of Innovation," *Research Policy* 6 (1977): 36–76, alluded, may be construed as a kind of metamarket.

54. The initial HYV wheat strain came out of an experiment that Norman Borlaug was conducting in Mexico in which he was moving test crops between high and low altitudes to gain two rotations per year. This experiment had the unexpected result of producing a strain that could grow well in a wide range of climatic conditions, contrary to established theories and experiences.

7

Contingency and Practical Reflexivity

TECHNOLOGY AND CONTINGENCY

Notwithstanding the antifoundationalist and antiessentialist attempts to deconstruct analytical philosophy in the wake of the recent historicist-post-positivist turn, the use-value of philosophical and historical concepts for the social study of technological change is still nontrivial.[1] Despite the contingency attributed to all knowledge claims and the rejection of methodological naturalism,[2] historicism does not overrule the possibility of observing some measure of regularity across time and space in social conditions related to (technological) knowledge production and dissemination. Philosophical exploration on technological change is appropriate because of the epistemological issues embedded in its conceptualization and explanation. The historiographic development in the history of technology to treat "technology as knowledge"[3] makes it relevant to apply concepts from the philosophy of science and technology to theorizing and explaining technological change. However, an epistemic relativism that resorts to a methodology of "anything goes" is rejected in theorizing and explaining technological change.

More importantly, conceptualizing and explaining technological change, as such, involves three important categorical issues. The first refers to the physical and material aspects of the artifacts involved, which require a causal-cum-functional explanation. The second involves the physical and biological constitution of the human (sometimes nonhuman) agents, which require a functional-cum-intentional explanation. And finally the agency question pertains to socialization, individuality, reflexivity, and so on, which requires an intentional explanation. The issue of social theory and hence the role of social and human sciences in explaining technological change enters the exercise through the third stage. However, even more, narrating the former two through social theories, language, and other tools of discourse makes social science important for explaining technological change.

The process of technological change is a contingently shaped historical process that involves a social phenomenon, namely the diffusion of technological knowledge in society. The objective of this chapter is to understand how the process of technology transfer actually takes place as a dimensional process involving space and time vectors. This is undertaken within the theoretical framework of technology as knowledge and technological change as knowledge change. This exercise, as we will see later, has important technology policy implications. The empirical evidence to verify the arguments is drawn from the case of the technology transfer in the Green Revolution in Indian agriculture.

PHILOSOPHY OF SCIENCE/TECHNOLOGY
AND THE MODES OF EXPLANATION

Besides the legitimation, justification, and verification of knowledge claims in all scientific, technical, social science, and humanities and arts disciplines, philosophical concepts are invoked to study the features that are common among all these disciplines. Of equal importance are the features and conventions that are usually invoked to separate the disciplines from one another. The two ways of classifying disciplines are (1) the particular method of legitimation, justification, and verification of knowledge claims, and (2) the particular mode of representation and explanation used in these disciplines for the above tasks. The characteristic methods of the natural sciences and technical disciplines, according to empiricist philosophers, are the hypothetico-deductive method of theory confirmation, or the deductive-nomological mode of explanation, which is also known as the covering law model of explanation.[4] In the arts and humanities disciplines, the method employed is the hermeneutic method. In the social sciences, the method of dialectics is usually the mode of explanation.[5]

This is only an approximate and obviously dated classification of disciplines according to "method." Rather, it would be a futile exercise to differentiate disciplines based on a single category of "method." There is no rule that stipulates that the dialectical method cannot be used in the arts and humanities, or the hermeneutical method and the method of hypothesis in the social sciences, or for that matter, these two methods applied in the "hard" natural sciences. Even stating that the hypothetico-deductive method is *the* method of explanation in the hard sciences is a position that can be easily deconstructed. Elster claims that the hypothetico-deductive method is *the* method for "all empirical sciences."[6] Then again, what counts as empirical sciences? Do social sciences fall under this category?[7] Popper correctly pointed out that despite the antinaturalistic beliefs of social scientists, most of whom are wedded to historicism, there are several common elements in the methods of physical and social sciences. It is also not quite clear which of the methods is conducive to theory building, in the wake of the historicist-nominalist turn discussed earlier.

Elster delineates three types of explanation for the sciences, including social sciences: (1) causal, (2) functional, and (3) intentional.[8] Accordingly, the best explanation applicable for physics, the quintessential "empirical science," is a causal explanation. Physical laws and phenomena can best be explained by tracing the causes for the particular phenomena or the events that the physical theories and experimental entities are supposed to explain.[9] Since functional explanation anticipates notions such as adaptation, selection, and evolution, it cannot be applicable to physics. So is an intentional explanation that can be applicable only for biological agents with subjectivity. Among the sciences, biology stands midway between physics and social science. In its case, the best explanatory mode is a functional one. It can be asserted that causal and intentional explanations are also applicable to biological sciences. For social sciences, obviously, the best mode of explanation is intentional. However, causal explanation is very much applicable to social sciences as well.[10] Human behavior and social changes need intentional and causal explanations to be meaningful. Only intentional beings are capable of adapting and behaving according to policy strictures and preplanned strategies. Elster however maintains that functional explanation has no role in the social sciences.[11] Finally, explanations should not be looked up on as an end in themselves. They can be often contingent and should not be viewed as immutable. Explanations can change as theories change, and also when we know more about the causal factors behind physical, natural, and social phenomena.

In the Humean view of causation, causal relationship can hold only between events as constant conjunction of one event following another. This claim should not be equated to a deterministic and mechanistic world. Contingencies and intentional and unintentional actions carried out by agents can also lead to "events." In fact, this antimetaphysical and positivist prohibition against belief in causes is not valid anymore as experimental methods have become adept at establishing the existence of numerous entities that were previously thought to be mere fictions or theoretical tools. In general, event X or agent X caused event Y: $X \Rightarrow Y$. A causal relationship may normally obey three laws: (1) determinism, (2) local causality, and (3) temporal asymmetry. Determinism simply means that every event should have a cause. The only instance in physical nature where this law is violated is when explaining the behavior of elementary particles in quantum mechanics, where explanation can only be statistical and probabilistic, as expounded by the uncertainty principle of Werner Heisenberg. Local causality means that action at a distance is impossible. To phrase it differently, "a cause always acts on what is contiguous to it, in space and time."[12] Phenomena that were assumed to defy local causality tend to display hysteresis, which means that there is a time lag between causes and events.[13] Finally, temporal asymmetry means that a cause must always precede its effect, or theoretically the cause must not succeed the event. However, in the microrealm of quantum mechanics, temporal asymmetry is rampant.

Causal explanations do have great importance in the social sciences because of hysteresis and counterfactual reasoning. Hysteresis is a significant factor in social sciences because of the importance of experience and history in social theory. Historical experience and time lag between cause and effect play more important roles in social science models and theories than in the physical sciences, "because the structural knowledge that would enable us to do away with the apparent hysteresis has not been attained."[14]

Counterfactual (or contrary-to-fact) argument is important in social science because of its normative dimension for inducing social change in a desirable way by appealing to a possible world, had certain actions occurred or been consciously carried out in the first place. The importance of counterfactual argument stems from the notion of causality (from a philosophical point of view) and the importance of evaluating the relative importance of causes (from a historical point of view).[15] Counterfactual argument's appeal is not merely to reflect on the past, but also to shape the present and the future by influencing present policies. A causal statement means that "event X caused event Y." On the other hand, a counterfactual statement means, "if event X had not occurred, event Y would not have occurred." Although counterfactual statements do not change causal statements, or the "truth" status or verities of actual historical events, the former do have a constructive role in social science. It becomes useful in weighting the relative causes of an event X, such that it had causes $C1, C2, \ldots, Cn$. If there is no indexing problem, the different causes can be weighted to a common measure and we can find out the relative contribution of each cause for the event X in a quantitative manner.[16]

Elster argues that epistemological difficulties may develop because of the inability to ascertain the truth value of a counterfactual statement, or even because of the problem of the assertability of counterfactual statements.[17] But this again is a vestige of teleology in social science thinking. Of course, counterfactual statements are related to the real world historical events. A counterfactual statement only argues that what if the technological innovation took a different form, given the possibilities of an alternative trajectory, from the actual innovation that actually happened. Given the possibility that Union Carbide had the technological wherewithal to produce a less-lethal pesticide given its enormous knowledge-base and expertise in chemical technology, instead of using methyl isocyanate (MIC), a chemical explosion in its Bhopal plant under the same circumstances, as in the 1984 (real) explosion, would not have killed 15,000 people (5,000 immediately after the explosion and 10,000 during the past fifteen years) and injured 300,000 residents of Bhopal.

As referred to earlier, intentional explanation takes a special position in social sciences because of the issues of behavior, beliefs, reflexivity, and agency. Intentional explanation essentially deals with the future by taking into account present and past information as a guiding light for theory building. Intentionality as purposive action to influence or change a future event brings in notions of rationality, according to Elster. But rationality of a bounded nature can only be defended, such as "local maximization," given the Kuhnian revolution in intellectual history, particularly the demystification of methodology and rationality in the construction of scientific knowledge. However, Elster rightly points out that rationality cannot be equated with optimality while explaining the intentional actions of the agents.

In explaining the case study of the Green Revolution and ascertaining the usefulness of practical reflexivity as a way of theorizing the technological change that has taken place in Indian agriculture, all the three explanatory modes will be invoked. However, the intentional mode will certainly be the most important one because we are trying to understand a technological change by looking through the contingently shaped events caused by the intentional actions of several agents. Before this exercise, let us first find out the boundary problems and the question of practical reflexivity.

BOUNDARY PROBLEMS, CONSTRUCTIVISM, AND PRACTICAL REFLEXIVITY

As argued above, philosophical parameters are invariably invoked when conceptualizing the process of technological change. Boundary and demarcation problems hamper the social study of science and technology,[18] and category-defining exercises are invariably invoked to mitigate and tidy up the problems. One major problem with theories of scientific change or technological change drawn from contemporary social theory or analytical philosophy, even in its historicist metamorphosis, is the problem of the separation of metatheoretical or second-order questions as distinct from theoretical or first-order questions. This problem is inevitable, given the metaphysical baggage that much of modern philosophy and social sciences have inherited. Although the antifoundationalist bath water may throw away the epistemological (including all theorizing activities) baby of all disciplines,[19] a pragmatist standpoint is taken to retain the need for theorizing to solve the problems that most social and human sciences had taken for granted as their central normative concern and disciplinary agenda.[20]

It is a nontrivial issue that if a first-order category were to undertake the justification of its own categories, then a theory couched in some sets of terms would have to validate these same terms thus creating a situation that ends up in the theory pulling itself up by its own bootstraps. This would obviously presuppose what it was supposed to show in the first place. Thus, a vicious circularity trap becomes the end result of theorizing. In simple terms, this trap is caused by the fact that all theories are representations. Any attempt to justify a representation would require another representation and so on, thus denying us a way to "hook up" to the real world. On the other hand, the *a priori* specification of a regime of metatheory (second-order condition) as distinct from theory (first-order condition) can save one from getting into this trap. However, this can only be achieved with fateful consequences. If a second-order (theory) category is invented to validate a first-order (empirical) category, we will need a third-order (metatheory) to validate the second-order, and so on, and so forth. The Scilla of vicious circularity is avoided, but for a Charybdis of infinite regress. However, this latter problem is very much overlooked or ignored because the validity and status of categories are construed as metatheoretical issues (hence, a nonstarter) by disallowing first-order types reflecting on categories. Thus, philosophical or social reflecting on real world empirical issues became metatheoretical in nature, and consequently making the project of "unmasking" the "real" hopelessly unattainable.[21]

The importance of the Kuhnian revolution in modern intellectual history was precisely its unintended historicist turn, pointing out the theory-ladenness of empirical data.[22] Kuhn's work was also a watershed in modern intellectual history by dismantling the positivists' "noble dream" of constructing "value-free" knowledge through a subject-independent and "objective" mode of inquiry.[23] However, in this process, Kuhn unwittingly succeeded in blunting the idealism of scientific practice as a critical enterprise, by reducing normal science to a mere problem-solving activity. The reduction of all intellectual practices in the social and natural sciences to a normalized problem-solving activity, without delving into the political-economic and normative dimensions of this activity, blunted the critical edge to cultural and social studies of science and technology. Although the Kuhnian revolution in modern intellectual history succeeded in solving the paradigmatic puzzles of disciplines and individual practitioners of the disciplines, Fuller correctly points out that the undermining of the dualism—theory-observation, fact-value, empirical-normative, and so on—in the practice of science, diminished the normative claims of the philosophy of science.[24]

One of the objectives of this chapter is to bring back the normative while remaining faithful to the nominalist-historicist beliefs of the practitioner.[25] This chapter will show a way to reinstate theorizing that is theoretical and metatheoretical at the same time. Overtly fulminating against the philosophical tradition of his time, Marx (following Hegel) meant to maintain this normative claim exactly by exhorting the philosophers that their objective should not merely be to understand the world, but to change it.[26] Marx conceptualized theory and practice (praxis) as two aspects of the totality of the unity of human (social) experience. That is, any theory should reflect or correspond to practice. Richard Gunn aptly characterizes this method. According to Gunn,

> Theory is *reflexive* when it reflects on the question of the validity of, or justification for its own categories. Theory is practically reflexive when in doing so it reflects upon and understands itself as inhering in a practical (a social) world. In essence, theorization should be based on practical (real world) experience.[27]

The reflexive practitioner should duly acknowledge the contingency and ephemerality of theories in such practices. All theories of social phenomena, including theories of technological change, should reflect the contextual practical realities of the political-economic structures and social relations of societies and nation states. This reflexive nature inherent in the theory of technological change (that technological change is knowledge change) will be shown with the help of the technology transfer in the Green Revolution in the context of the agricultural and economic modernization practices of postindependent India. In preparation, let us see how the problem of practical reflexivity and the construction of theories from empirical evidence or facts was addressed by a great historicist philosopher of science (William Whewell), and by one of the greatest practitioners of science (Charles Darwin), both from the nineteenth century.

The ability to form theories and laws about the causal linkages between events is a reflection of the maturity of practice and growth of knowledge in the "empirical" sciences. Formulating theories from empirical observations is a valuable and legitimate method to understand any phenomena. The Baconian inductivists, on the other hand, held that "facts" arranged in a "logical" way would reveal their inner secrets, and consequently to theories and laws to explain the "facts." However, what is most important in the history of modern intellectual thought is the realization that there is no subject-independent and objective way of knowing, and consequently of knowledge production itself as Kuhn and other historicists have shown. This Baconian method is more of a "fiction" as the historicist philosopher of science William Whewell showed, though ironically, Whewell himself characterized theorizing in science as an inductive process.[28] Whewell argued that in order for making sense of the disorder and chaos of the given "facts" of the world, the observer's subjectivity has to be imputed.[29] That is, the observer has to impose some general "ideas" to make sense of the "given facts." For Whewell, facts are the material foundations of science, and science begins with the common observation of facts, and all facts involve ideas.[30] We can presume to know something about the phenomena about which facts are being collected by framing plausible hypotheses, and then selecting the right hypothesis among the plausible ones.[31] In order for the theories to have predictive or explanatory value, one needs to employ the appropriate test method or explanatory mode.

Darwin was the quintessential scientist who resorted to this practical-reflexive method of theorizing from empirical data (observation), or what Whewell calls "facts." Darwin's bold conjecturing and speculative approach is reflected in the opening paragraph of *On the Origin of Species:*

> When on board H. M. S. 'Beagle,' as naturalist, I was much struck with certain facts in the distribution of the inhabitants of South America, and in the geological relations of the present to the past inhabitants of that continent. These facts seemed to me to throw some light on the origin of species—that mystery of mysteries, as it has been called by one of our greatest philosophers. On my return home, it occurred to me, in 1837, that something might perhaps be made out on this question by patiently accumulating and reflecting on all sorts of *facts* which could possibly have any bearing on it. After five years' work I allowed myself to speculate on the subject.[32]

Darwin always retained that one can only observe through the lens of a theory.[33] Darwin affirmed this theory-ladenness of observation in one of his letters of 1861 in which he wrote:

About thirty years ago, there was much talk that geologists ought only to observe and
not theorize; and I well remember someone saying that at this rate a man might as well
go into a gravel-pit and count the pebbles and describe the colours. How odd it is that
anyone should not see that all observation must be for or against some view if it is to
be of any service![34]

Darwin's theory of evolution postulating natural selection as the causal explanation for the evolution of species is one of the clearest examples of practical-
reflexive theorizing in science.

However, as Popper astutely pointed out, science does not begin, nor does
knowledge grow with observations or by mere collection of data.[35] All great
scientists like Darwin certainly knew this stricture. What is required is the
Whewellian idea of looking for a certain *kind* of data. According to Popper:

Before we can collect data, our interest in *data of a certain kind* must be aroused: the
problem always comes first. The problem in its turn may be suggested by practical
needs, or by scientific or pre-scientific beliefs which, for some reason or other, appear
to be in need of revision.[36]

As explained earlier, the particular problem, for example the origin of species,
arises because of the need for certain explanation of phenomena, regularity, variation, and so on. The "logic" of scientific discovery is to explain, preferably
through hypotheses and theories, the problem we encounter in understanding phenomena, social or natural. Thus, theories cannot be proved true or false by verify-
ing the truth status of statements; they can only be corroborated through better
evidence.[37] Essentially, the *method* is to offer deducible causal explanations, predictions, and tests; and there cannot be any absolute certainty for any of the theories expounded to capture the relationship between causes and events. However,
one needs to concede to Hacking's claim that we can show some theories do not
"fit" or mesh with nature.[38] However, that does not hold that there is a correspondence theory of truth, or that science will converge toward an ultimate "truth," despite the realist position that there is an independent world (of the agent) out there.

PRACTICAL REFLEXIVITY AND TECHNOLOGICAL CHANGE

The most significant theoretical contribution to conceptualizing technological
change evolved as a result of the recognition of the epistemic significance of technology as knowledge in the historiography of technology. The theoretical problem
of accounting for the growth of knowledge in the history of technology became
crucial, as technology was not seen as a handmaiden of science. The ahistorical
assumption that technology is merely applied science has been long abandoned.
As the intellectual autonomy of technology as a distinct knowledge system, separate from science, began to be recognized, modes of theorizing and explaining

technological change as a process distinct from scientific change emerged as a significant problem in science and technology studies.

The epistemic focus of technological practice is "doing" and "solving" practical problems, and the corresponding focus for scientific practice is "knowing." Theoretical reflection on the enterprise of technology did not evolve because of the ascription of technology as constituted by its artifactual realm. Theories and models of technology transfer and diffusion failed to deliver because of the emphasis given to the physical constitution of technology rather than its social and cognitive context and content. Any meaningful analysis of technology transfer should be undertaken by transcending the artifactual realm of the techniques and then shifting the analytic focus onto the realm of the actors involved at both ends of the technological knowledge transaction (production and diffusion). When technology transfer is undertaken, either voluntarily or otherwise, the diffusion of the new technological knowledge brings about technological change. It is a slowly evolving, cumulative, and continuous process. It involves learning by observing, learning by doing, learning by scaling, learning by tinkering, and other extant processes.

The significance of technological change as knowledge change stems from the triangular interaction between technology, culture, and society. The transferred or indigenously evolved technological knowledge is the result of changes in the means of production of the dominant actors and social groups. These actors can be firms, entrepreneurs, governments, and farmers, among others. In the case of agriculture, through the agency of the government and its extension and outreach services the new knowledge is diffused to the farmers and peasants. Technological change is not merely an anonymous accumulation of artifacts, but the growth of knowledge related to social production and reproduction.

The knowledge change that took place in the Green Revolution can be explained as an intentional process that was based on a package of policy recommendations that was laid out in advance to influence and change the future behavior and actions of selected human actors as well as artifacts in a "rational" way. The Green Revolution is generally referred to as the increase in cereal productivity that India and several other Third World countries experienced in the 1970s as a result of the modernization of agricultural technology and practices to boost food production. An explanation of the historical contingency of how this process acquired a name like the Green Revolution and a comprehensive historical reconstruction of the technological change in Indian agriculture was attempted elsewhere[39] and in the previous chapter.

After independence, the leaders of the newly founded Indian republic wanted a planned restructuring to rejuvenate the Indian economy and industry that they inherited from the British colonial administration. The leaders wanted to achieve these goals by resorting to massive industrial development campaigns. Agricultural modernization, however, was not given high priority when the industrialization drive was being planned. The implicit objective was to redirect the surplus from the agricultural sector to the industrial sector (in the form of underpriced and subsidized food for the industrial workforce and urban dwellers) as happened in the Soviet Union during Stalin's industrialization drive in the 1930s and 1940s.[40] The neglect of the agricultural sector and several crop failures due to tardy monsoons created tremendous shortages of foodstuffs in India in the mid-1960s leading to near-famine conditions. The food shortages were temporarily alleviated by massive shipments of cereals from the United States and other countries. In order to increase cereal production, the Indian government, under pressure from the aid donors and the World Bank, decided to implement a package of agricultural modernization policies, the result of which was dubbed the Green Revolution.

The Green Revolution involved the use of wheat and rice seeds of high-yielding varieties (HYVs), which were developed first in Mexico and the Philippines, respectively, and the concomitant adoption of a package of new agricultural practices involving chemical fertilizers, pesticides, irrigated water, and some mechanized implements. No radical changes in agrarian relations were implemented, such as land reforms and the distribution of surplus lands to landless peasants. The policy was primarily aimed at increasing agricultural productivity by changing the factor endowments of small and medium landholding farmers and peasants. The Indian government's objective was to procure the extra production, after meeting the subsistence needs of the farmers and peasants, at a rate determined by the Agriculture Price Commission. The spread of the Green Revolution was uneven. The effect of the change was felt only in a few states such as in the Punjab and Haryana and in parts of Karnataka, Utter Pradesh, and Tamil Nadu. Although the Green Revolution affected less than a quarter of the total land area under cultivation, the increase in productivity made India self-sufficient in grain production.[41]

The spatiotemporal spread of the Green Revolution shows some distinct diffusion patterns and stages spread over a period of few decades. The first stage is the development of a new and indigenous national and state level agricultural research system involving national agricultural research institutes and agricultural universities that were patterned after the land-grant state universities in the United States. The second stage is highlighted by the overhaul and reform of the agricultural bureaucracy in India to facilitate the transfer and diffusion of the high-yielding varieties of seeds and the establishment of an agricultural extension system. The third stage is marked by the change in agricultural practices of the adopters as a result of the introduction of the HYVs and the diffusion of the new agricultural knowledge.

Some features of the United States agricultural research and extension model were replicated in India during the period through the state governments and the state agricultural universities. A fairly sophisticated and location-specific national agricultural research capacity was established as a result of the policy package. The existence of a highly educated elite (many of them trained in the United States and Britain) once convinced of new opportunities in the agricultural sector helped to establish the national agricultural research system. The transfer process in the Green Revolution involved not only the seeds and other physical artifacts, but also the knowledge that returning scientists brought as well as the knowledge-producing capability that the agricultural research system entailed. The development of a sophisticated, national agricultural research capacity made it possible to adapt the new knowledge to the specific needs and conditions of the farmers and peasants who were willing to participate in the new agricultural programs. The government agencies worked out an elaborate incentive scheme to attract the farmers and peasants to the HYV-based agricultural practice. The peasants and farmers learned the new agricultural practices by observing the extension agents, through trial-and-error, and other such interactive learning processes.

The Green Revolution shows that the technology transfer process entails novel ways and means of practising agriculture. It shows that the traditional practices and expectations changed to a new pattern of buying seeds and inputs from the market in exchange for the goods the farmers produced, thus blurring and expanding boundaries of the sociotechnological system. In the old system, the boundary of the sociotechnological system did not stretch beyond the villages. With the new system, it stretched far beyond the villages. As a result of the dissemination of the new knowledge of agriculture through increased communication facilities, the sociotechnological system expanded to include markets, rural cooperatives, roads, irrigation systems, extension networks, national and international research and educational institutions, new seeds, pesticides, different government agencies, and so on. The change from the "traditional" ways to the new ways was not instantaneous, but gradual. The peasants and farmers experimented with the HYVs by first devoting only a small fraction of their land to them. The learning process involved feedback between extension agents, scientists, and farmers. The HYVs had to have their genetic profile adapted to Indian soil and climatic conditions and culinary habits. Field research shows that the peasants accepted the HYVs and the associated technological package only after they were convinced of the high-yielding potential.

In both the new and old systems, the technologies reflect the ideas and experiences of the actors involved. Both systems were efficient when viewed from their own bounded rational positions. In terms of total energy consumption versus output, the old system was more efficient than its modern counterpart. The development of the HYVs show that they were the result of a knowledge system that evolved out of rich human experiences dating back thousands of years. The basic idea behind its development was energy efficiency. A smaller stalk meant less energy wasted in the stalk. But its full potential emerged only in the late nineteenth and early twentieth centuries when chemical fertilizer was invented. The new package of agricultural practices associated with the HYVs was clearly the result of an instrumental ordering of human experiences and ideas with efficient means as the fundamental principle of all learning processes.

The network of causal linkages in the Green Revolution can be put as follows:

invention (hybrid seeds, HYVs, seed-fertilizer-irrigation agriculture)⇒transfer (of the new agricultural practice and the HYVs)⇒innovation (further adaptive research in India to make the HYVs location-specific)⇒development (establishment of research capability, development of several other new varieties)⇒diffusion (the transmission of the new knowledge to the farmers and peasants through extension agents) ⇒feedback (problems and new ideas relayed back to the researchers through extension agents)⇒innovation (further research and development)⇒transfer (of the new technological knowledge to other parts of India).

This is an open-ended network. The interactive pattern may be much more complex than this unidirectional flow chart shows. These stages and factors acted interactively without any hierarchical structures.

The Green Revolution shows the evolution of an algorithm of the way in which new ideas and information are transformed into material output in the form of food. The algorithm may be construed as a means for simplifying the complex knowledge of the agricultural experts into simplified steps that the peasants and farmers could adopt into their practices. The process of the change was conceptual in terms of the change in cognition of the agents (actors). The technological change that is characterized as the Green Revolution involved not just the introduction of a few new material or tangible artifacts, but the change in the knowledge of the practitioners. Theories or models of technological change should emphasize how the knowledge content of technology changes, and hence emphasis should be placed on the actors and their social and institutional contexts and structures. The appearance of newer artifacts (HYVs, chemical fertilizers, electric and diesel pumps, pesticides, irrigation canals, mechanically powered agricultural implements, and so on) should be looked upon as the physical manifestation of technological change at the secondary or tertiary level. Thus, the practically reflexive theoretical claim that technological change is primarily a process of knowledge change of the actors is being affirmed.

It is a fact that the Green Revolution involved a change in the practice of agriculture in parts of India. The heterogeneous network of the elements of agriculture technology shows the enrollment of new actors while several old actors disappeared. The knowledge change of one of the significant actors, the peasants and farmers, forms the core link in the network as well as providing the key problem as Popper would put it. Several field researchers reported the change. Following the Darwinian-Whewellian methodology, the best mode for explaining the change is an intentional-cum-functional one. From this, we can speculate that the technological change in the case of the Green Revolution was essentially a knowledge change of the actors.

PROBLEMATIZING THEORY AND THE GROWTH OF KNOWLEDGE

In general terms, conceptualization of technological change is problematic enough. At a more special level, a clear understanding of the dynamics involved in the Green Revolution presents an even greater challenge. It is not merely the transfer of *a* technology from one country to another, nor the transfer of *an* artifact from the laboratory to the field. Based on reflexive analysis, the theoretical claim that the technological change that took place here represents knowledge change could provide a tentative solution to the problem raised above. Thus, following Popper we can look upon any knowledge claim and especially upon any theory "as a tentative solution to some problem or other, and as giving rise to new problems. And the fertility and depth of our theories may well be measured by the fertility and depth of the new problems to which they give rise."[42] The problem that a theory is intended to solve can be a practical problem, or another theoretical problem such as explaining phenomena, social or natural.[43] Practical problems can give rise to theoretical problems, and vice versa. No claim is made that my theoretical attempt here solves all the problems of explaining the Green Revolution, an empirical problem. My tentative theoretical proposition is that it is a critical attempt to solve the problem of the growth of knowledge in the process of technological change.

The application of the evolutionary metamodel is only tangential in scope in conceptualizing technological change. In the case of the Green Revolution, the government and its agencies helped to create a selection environment for the actors-cum-practitioners. Unlike in natural selection where random processes determine the outcome, persuasion through propaganda, incentive schemes, and unanticipated and contingent factors decided the outcome of the selection process. The possibilities for variation are limited and fixed *ex ante* by the government and the market forces because of the pre-established intention of changing the behavior of the actors. It shows that the public agencies can guide the selection process, to a large extent. Expected and unintended consequences can always upset the outcome. Some factors are the Bhopal gas explosion, structural changes in Indian agriculture, and the high incidence of accidents and deaths due to inadequate training of the agricultural workers who lacked protective gears and clothing while handling chemical pesticides and dangerous mechanical implements like weeders and threshers.[44] Adverse and unexpected effects due to the new selection may in turn cause a change in policies.

It behooves us to open the black box of technological change in order to see it as an endogenous process involving not only the changes in the physical constitution of the tools and things, but the essential way of "mixing things" to create new things and to produce old things efficiently, cheaply, and abundantly.[45] Technological change as knowledge change is predicated upon the antideterministic claim that the practitioners determine the locus or trajectory of technological change, and not any "internal logic" of the technology. Although there is no noncircular way of affirming that technology is socially constructed, following the actors through society and enrolling in their network is the only way to understand their practices and knowledge claims.[46] There is no referent and reference outside the actors and their (social) world on which to reflect and, subsequently, to theorize. We are grounded at a place and all of us can claim a place-based identity.[47] Any claim of an independent framework for ascertaining the growth of knowledge is a myth.

ENDNOTES

Some of the material in this chapter is drawn from my article, "Practical Reflexivity as a Heuristic for Theorizing Technological Change," *Technology in Society* 19, no. 2 (1997): 161–175. Acknowledgment is duly accorded to Elsevier Science for the necessary permissions.

1. The historicist turn in intellectual history can be traced to Hegel, on whose trail Marx, Mill, Nietzsche, Heidegger, Dewey, Derrida, Foucault, and Rorty, among others, added radical contingency to human knowledge claims.

2. Unlike the possibility of generalization as a typical observational or explanatory concept in the natural and physical sciences, such a feature cannot be attributed to social conditions, past and present, among all social groups. Also, a social group is not merely the physical agglomeration of the members of the group. The rejection of methodological naturalism may entail the acceptance of nominalism as a universal given in the social sciences. However, nominalism is retained only to reject essentialism in the Aristotelian sense. Social sciences need to retain some form of methodological essentialism to understand the "essence" of social phenomena, sans any teleological pretensions, in order to establish the causes behind these events. One cannot understand colonialism or slavery by a mere nominalist account of its horrors alone. One needs to go further than mere description of social phenomenon such as imperialism to understand the root causes. The success of physical sciences may be attributed to methodological naturalism and instrumentalism. Despite the rejection of essentialism as an ideology, the social sciences ironically need some form of methodological essentialism to retain their normative claims.

3. Edwin T. Layton, "Technology as Knowledge," *Technology and Culture* 15, no. 1 (1974): 31–41.

4. The hypothetico-deductive method (or the method of hypothesis) may be simply put as the method of offering deductive causal explanations and testing them through predictions. It is a method of theory confirmation by deducing from the hypothesis under test predictions whose truth or falsity can be ascertained through observations. According to Carl G. Hempel, *Philosophy of Natural Sciences* (Englewood Cliffs, NJ: Prentice-Hall, 1966), besides the deductive-nomological explanation, scientific phenomena can also be explained probabilistically in the covering law model of explanation. According to Karl Popper, *The Poverty of Historicism* (London: Ark Paperbacks, 1986), 131, the method of deduction "does not achieve absolute certainty for any of the scientific statements which it tests; rather, these statements always retain the character of tentative hypotheses, even though their character of tentativeness may cease to be obvious after they have passed a great number of severe tests."

5. See Jon Elster, *Explaining Technical Change: A Case Study in the Philosophy of Science* (Cambridge: Cambridge University Press, and Oslo: Universitetsforlaget, 1983), for further elaboration of this classificatory system.

6. Elster, *Technical Change,* 15.

7. According to Popper, *Poverty of Historicism,* any discipline that considers itself a branch of knowledge, which aims to be theoretical and empirical, falls under this category, physics and sociology being no exception to this general guideline.

8. Elster, *Technical Change.*

9. Elster, *Technical Change,* 18, certainly adds the caveat that the physics he has in mind is the classical, prerelativistic, and pre-quantum-theoretical physics.

10. One proviso is that in the social sciences one cannot resort to mathematically formulated causal laws to explain social phenomena, despite the utility of certain quantitative methods like statistics for explaining certain regularities (say in certain economic laws such as supply and demand in a bounded rationalist sense).

11. Elster, *Technical Change,* 20. Elster rejects functional explanation for social sciences on the ground that social events need not have any meanings. Attaching a meaning to social events is metaphysical and teleological in nature. Only theologians probably attach meanings to social events.

12. Elster, *Technical Change,* 28.

13. Magnetism and elasticity are classic examples of phenomena that exhibit hysteresis. Newton's theory of gravitation appeared to have "defied" local causality for the religious authorities, as the event (gravitational pull) did not seem to have a causal connection.

14. Elster, *Technical Change,* 33.

15. Elster, *Technical Change,* 175.

16. It should be reiterated that an "event" is an expansive concept in the social sciences that includes individual actions and behaviors that become the causes for events.

17. Elster, *Technical Change,* 36.

18. Despite Thomas Kuhn's, *The Structure of Scientific Revolutions* (Chicago: University of Chicago Press, 1970), denial that his model of scientific change through paradigm shifts did not apply to human and social sciences, it was in these disciplines that Kuhn's ideas made a lasting impact. However, Karl Popper, *The Logic of Scientific Discovery* (London: Routledge, 1992), laid down an explicitly rationalist account of demarcating "science" from "nonscience." Popper proposed the concept of "falsifiability" of hypotheses and theories as the demarcation criteria. Also, what distinguishes science from nonscience is the fallibility of all knowledge claims in the former, while the same cannot be said of the latter, such as religion and metaphysics where blind faith and unquestioned belief form the core of knowledge claims. For an interesting account of boundary work in the sociology of science, see Thomas Gieryn, "Boundary Work and the Demarcation of Science from Nonscience," *American Sociological Review* 48 (1983): 781–795.

19. Again, "epistemology" is taken in a minimalist sense, without any of the Aristotelian metaphysics of deriving knowledge from first principles. Also, it is not used in its classical or positivist incarnation to mean a systematic theory of knowledge based on explicitly stated rational methods and procedures, such as *a priori* methods of testing and verification of knowledge and theory claims, adherence to a correspondence theory of truth, and the assumption of referents outside human agents to avoid problems of subjectivity. On the contrary, it is used in a minimalist and pragmatist sense of a fund of useful human knowledge, of course, with any normative claims intact. Again, the minimalist epistemology that is invoked here is influenced by the evolutionary epistemology of Popper, *Scientific Discovery*. For an excellent exposition of Popper's evolutionary epistemology, see Donald Campbell, "Evolutionary Epistemology," in *The Philosophy of Karl Popper,* Paul A. Schlipp, ed. (La Salle, IL: Open Court, 1974): 413–463.

20. Foucault's attack on the formation of modern disciplines as a means of creating specialized knowledge as antithetical to individual autonomy, self-creation, and the collusion he saw between power and knowledge is very much based on this self-appointed agenda of modern social and human sciences.

21. This is in addition to the largely legitimate complaints raised against some meta-theories and metanarratives by postmodernists and poststructuralists, notably, Jean-

Francois Lyotard, *The Postmodern Condition: A Report on Knowledge* (Minneapolis, MN: University of Minnesota Press, 1984). For Lyotard, the "postmodern condition" is characterized by the "incredulity" toward all metanarratives.

22. Writing on the occasion of the thirtieth anniversary of the first publication of Thomas Kuhn's *The Structure of Scientific Revolutions* (1962), Steve Fuller, "Being There With Thomas Kuhn: A Parable for Postmodern Times," *History and Theory* 31 (1992): 241–275, astutely observes that Kuhn set out on his project (of accounting for scientific progress and change) with positivist intentions, and his historicist turn was almost serendipitous. According to Fuller, Kuhn wanted to leave the impression that he was more influenced by positivist, or at least more broadly "analytic," considerations about the nature of language and knowledge than by the historicist ones that his work supposedly generated.

23. For an excellent deconstruction of the objectivity dream in constructing value-free knowledge in the historical profession, see, Peter Novick, *That Noble Dream: The "Objectivity Question" and the American Historical Profession* (Cambridge and New York: Cambridge University Press, 1988).

24. Fuller, "Being There."

25. For an antiessentialist sketch of such a realist pragmatist, see Ian Hacking, *Representing and Intervening: Introductory Topics in the Philosophy of Science* (Cambridge, UK: Cambridge University Press, 1983), and for an antirealist pragmatist, see Richard Rorty, *Objectivism, Relativism, and Truth* (Cambridge: Cambridge University Press, 1991), *Contingency, Irony, and Solidarity* (Cambridge: Cambridge University Press, 1989), and *Consequences of Pragmatism: Essays, 1972–1980* (Minneapolis, MN: University of Minnesota Press, 1982).

26. Marx's famous eleventh thesis on Feuerbach is: "The philosophers have only interpreted the world in various ways; the point, however, is to change it."

27. Richard Gunn, "Marxism and Philosophy: A Critique of Critical Realism," *Capital & Class,* no. 37 (1988): 87–116, 87, emphasis in original.

28. William Whewell, *The Philosophy of the Inductive Sciences,* Volume I "Of Ideas in General," and Volume II "Of Knowledge" (New York and London: Johnson Reprint Corporation, 1967).

29. A variation of this concept is what Ilya Prigogine has shown in his nonlinear dynamics (chaos theory) that general features of chaotic behavior can be predicted by the established methods of the field. This is popularly known as the theory of self-organizing.

30. Whewell, *Inductive Sciences,* Vol. II, 467.

31. Whewell's, *Inductive Sciences,* Vol. II, 468, tenth aphorism concerning science is: "The process of scientific discovery is cautious and rigorous, not by abstaining from hypotheses, but by rigorously comparing hypotheses with facts, and by resolutely rejecting all which the comparison does not confirm."

32. Charles Darwin, *On the Origin of Species,* A Facsimile of the First Edition (Cambridge, MA: Harvard University Press, 1964), 1.

33. Ian Hacking, *Representing,* holds that this cherished view of scientists is a myth as observations can be made without holding on to any preconceived theories. According to Hacking, this notion that observation must precede theory is due to the historical bias against experimenters in the annals of science. According to him, it was just that the theoreticians became the Brahmins who determined the hierarchy of a scientific caste system in which the theoretician Brahmins put themselves at the top of the hierarchy of the scientific establishment.

34. Darwin quoted in Novick, *Noble Dream,* 35.

35. Popper, *Poverty of Historicism.*

36. Popper, *Poverty of Historicism,* 121.

37. Popper, *Scientific Discovery.*

38. Hacking, *Representing.*

39. Govindan Parayil, "The Green Revolution in India: A Case Study of Technological Change," *Technology and Culture* 33, no. 4 (1992): 737–756.

40. The Soviet planning model was adapted to the Indian situation in a "mixed" economic framework, in the form of five-year plans.

41. However, millions of landless peasants and agricultural workers went hungry during the Green Revolution (and still do) due to their inability to garner enough food at the marketplace because of their acute poverty and unemployment.

42. Karl Popper, *The Myth of the Framework: In Defence of Science and Rationality* (London: Routledge, 1994), 156.

43. Popper, *Framework,* 157.

44. The Union Carbide plant at Bhopal, which exploded in 1984 spewing tons of methyl isocyanate (MIC) that killed more than 15,000 people and injured 300,000 people, was manufacturing the pesticides for the Green Revolution. For details, see Govindan Parayil, "The 'Revealing' and 'Concealing' of Technology," *Southeast Asian Journal of Social Science* 26, no. 1 (1998): 17–28. Thousands of agricultural workers get killed and injured every year as a result of handling chemical pesticides and modern agricultural implements. According to William Bogard, *The Bhopal Tragedy: Language, Logic, and Politics in the Production of a Hazard* (Boulder, CO: Westview Press, 1989), half a million people are poisoned each year of which 22,000 die.

45. For an elaboration of "mixing" things as a metaphor for creating new things and technological artifacts, see Paul Romer, "Idea Gaps and Object Gaps in Economic Development," *Journal of Monetary Economics* 32 (1993): 543–573.

46. See chapter three for details of this social constructivist argument, and specifically Bruno Latour, *Science in Action: How to Follow Scientists and Engineers Through Society* (Cambridge, MA: Harvard University Press, 1987).

47. For an excellent place-based social theorizing, see Arif Dirlik, "Place-based Imagination: Globalism and the Politics of Place," *Review* 22, no. 2 (1999): 151–188.

8

Competing Models
and Their Explanatory Power

Several different models and theories of technological change were analyzed in chapters two through five with an objective to explore their possibilities and problems in explaining this phenomenon. As argued in chapter seven, an objective of any theory or model is to explain a phenomenon, social or natural. Thus the success and usefulness of models of technological change depend upon their ability to account for a social phenomenon that is caused by the changes in the material conditions and knowledge base of societies that are manifested by the changes in the constitution of relevant artifacts associated with the technological system. Let us look at how some of these prominent models (two models from each chapter are selected) might explain the technological change in the case of the Green Revolution, which was presented in detail in chapters six and seven.

HISTORICAL MODELS

Cumulative Synthesis Model

In this section, the cumulative synthesis model of Abbott Payson Usher and the systems model of Thomas Hughes, from the history of technology framework, will be used to explain the technological change in the Green Revolution. The "hard core" of the Usherian program is the theory of cumulative synthesis. That is, great technological developments are the cumulative synthesis of a large number of minor achievements. Usher does not subscribe to the notion of technological change as a disruptive and discontinuous process. Instead, he maintains that technological change is the end result of several minor inventions and other procedural technological activities cumulating over a period of time.

There are four steps involved in the inventive process, which Usher uses as a portmanteau for different technological activities that cumulate to form technological change. These four steps are (1) perception of a problem, (2) setting the stage, (3) the act of insight, and (4) critical revision. According to Usher, technological change is an evolutionary process involving minor and major (strategic) inventions. This processual approach is particularly appealing when explaining the development of the high-yielding varieties (HYVs) of rice and wheat seeds, although it is difficult to delineate the four stages specifically.

The "invention" of the HYVs was a great biotechnological achievement that can be conceptualized as a process of cumulative synthesis involving several small and large steps. Farmers knew for a long time how to develop these semidwarf varieties of rice and wheat through the biotechnological knowledge of crossbreeding.[1] However, it was the Japanese who began to exploit the dwarfness of the varieties in the 1900s for higher cereal productivity by applying chemical fertilizers. The developments of earlier HYV strains were to be credited to Japanese researchers who manipulated the genetic characteristics by crossbreeding them with local and foreign varieties. One of these successful wheat varieties, Norin 10, was introduced in the United States from where the Norin-Brevor variety was taken to the International Maize and Wheat Improvement Centre (CIMMYT) in Mexico. The genetic makeup of the modern HYV wheat was obtained from the Norin-Brevor by the efforts of Norman Borlaug and his colleagues at CIMMYT.[2] Thus, the development of the modern HYV wheat may be characterized as "an invention . . . that represents a substantial synthesis of old knowledge with new acts of insight."[3] Different rice varieties were developed, following the same technique as in the case of the wheat varieties, at the International Rice Research Institute (IRRI) in the Philippines. The IR-8, the most successful early rice variety, was obtained by manipulating the rice varieties (both dwarf and "regular" types) from Taiwan, China, and Indonesia.

The HYVs are one element of the Green Revolution network. The cumulative synthesis model may be able to explain the development of other components such as tractors, chemical fertilizers, pumps, modern irrigation systems, and so on. However, a disaggregated account of the various components can not tell us much about the process of technological change taken place here. If we disregard the four-stage processual model, and instead take the theory of cumulative synthesis as a general explanatory model of technological change, we can explain the Green Revolution as the outcome of a process made up of numerous steps. The changing practice of agriculture during the period from its inception in the early 1960s to the 1970s may be taken as an evolutionary process that culminated in the technological change. This process includes steps such as invention, transfer, diffusion, development, and innovation, not necessarily in any rigid order. Although the model explains certain elements of the network, it does not help us account for the intangible and disembodied aspects of the Green Revolution. The technological change that has taken place in the Green Revolution can be accurately accounted for by looking at the process of knowledge change of the agents, the peasant-farmers, as shown in great detail in chapters six and seven.

Systems Model

Thomas Hughes shows that the historical development of all large-scale technologies can be conceptualized in systemic terms. As we saw earlier in chapter two, in the Hughes model the process of technological change is analyzed by systematically unveiling the five phases or stages of development of the particular technological system. These five phases are the beginning of the system, usually by an inventor-entrepreneur; the technology transfer and developmental stage of the system; the growth of the system during which serious problems (reverse salients) affecting system growth are solved; the acquisition of momentum during which the system gains direction and velocity; and finally the mature stage of the system during which the system may enter into a permanent state of stasis or begin to decline and disintegrate.

The conceptualization of technological change in the Green Revolution, according to Hughes's model, would be the growth of peasant agriculture dominated by "traditional" technologies and its development into near-commercial (influenced by market factors) farming methods dominated by modern technologies. However, the peasant agriculture system dominated by traditional inputs like traditional seeds (whose genetic makeup went back hundreds of years), wooden plows, rain-fed water, and manure and the modern seed-fertilizer-controlled-irrigation agriculture system follow different dynamics of system growth and performance. The important factor that drives peasant agriculture can be characterized as subsistence farming that carries low risk. Although the productivity of the land is low, peasants can expect a subsistence crop every season, unless major natural disasters strike. On the other hand, modern HYV-based farming is largely predicated on the concept of risk taking (influenced by the behavior of market and nature), although one of the most important functions of the technological inputs is to reduce the amount of risk by the process of "learning by doing," "learning by using," and other methods of innovation. Although functionally the same in terms of the nature of the output, traditional and modern agriculture are qualitatively different processes. The obvious trade-off is higher productivity of the land in the latter case, with entirely unique system dynamics.

Hughes does not provide a clear explanation as to how the system's boundaries change during the growth process of the system. If the process is the expansion of the system from traditional to the modern in the case of the Green Revolution following Hughes's model, it is not clear how the new components can fit in with the old system or how the new system replaces the old one. System growth is considered a "seamless web" of different components networked together while seemingly intractable problems are solved by enterprising manager-entrepreneurs. Since the agricultural modernization program was orchestrated by public agencies (governments), bilateral and multilateral aid agencies, international agricultural research institutes, and so on, it is difficult to locate manager-entrepreneurs, although there were several key scientists and bureaucrats involved in problem solving. Also, during the beginning of the system, identifying a key individual as the inventor-entrepreneur who laid the foundation for the system would be difficult.

In the Green Revolution, the technology transfer phase of this model applies better than the other four phases. The modality of technology transfer that Hughes delineates is one of importing new technologies to the recipient nation or region and soliciting capital from local sources for further development and consolidation. Patent rights in the host country protect the new imports. But the host country's legal and economic contexts and constraints determined the diffusion of the new technology and the system growth. Although there were no intellectual property rights issues involved during the early stages of the Green Revolution, the transfer of research capacity was regulated by political, economic, and related institutional constraints. In the case of the Green Revolution, the "investors" were governments and donor agencies, where profit was not the major "investment" criterion.

The third phase of Hughes's model is concerned with system growth during which potential reverse salients that crop up during the system growth have to be solved. A major reverse salient in the agricultural situation was declining productivity due to droughts and stasis in technological change. In a certain sense, we can suggest that the replacement of most traditional inputs with their modern counterpart was the action that reversed the critical problem created by the declining land productivity. The need to make the new seeds location-specific, to deal with issues of nature (pests, droughts, etc.) and social mores (tastes, preference for certain hues of the cereals, etc.) was a daunting problem that the planners and scientists had to deal with. However, the newly created research capability was able to solve this problem successfully. Making agricultural implements like small tractors and locally designed pumps were all part of this effort.

The fourth phase of the system growth is momentum. In this sense the "mass" of the system include the physical artifacts (seeds, pumps, irrigation systems, fertilizers, tractors, and so forth), institutions (universities, research centers, extension networks), and other such entities associated with the Green Revolution. The "velocity" of the system may be construed as the rate at which the new knowledge was diffused and accepted by the peasant-farmers. The "direction" of the system can be subsumed under the policy directives of the government and other public agencies to increase productivity by inducing changes in agricultural practices. Abstracting these concepts from the case study can be a difficult process and hence should be construed in a metaphorical sense.

Finally, it is rather difficult to find a corresponding fifth phase of the system in the case of the Green Revolution. One can detect signs of stasis and decline in productivity. However, agriculture, as a human activity, will remain until food can be synthetically produced in factories, instead of on land.

The systems model can be used to explain the Green Revolution only if we take extreme liberty with what constitute the boundary conditions of the system. Agriculture, the first organized human enterprise, is a rather difficult entity to be brought under the parameters of a system. It may be even safer to treat major components of the Green Revolution as separate systems. Notwithstanding these problems, Hughes's model can be applied to the Green Revolution provided we take extreme liberty in interpreting the model. We should concede that Hughes's model of system evolution was originally developed for a sociotechnological system that was more localized and found largely in advanced industrialized economies. Most of the institutional parameters Hughes takes for granted are rather rudimentary or nonexistent in India during the Green Revolution. My earlier concern still remains as to whether the systems model of Hughes can be applied to other systems with a different dynamics of growth and development.

SOCIOLOGICAL MODELS

Social Construction of Technology (SCOT) Model

The conceptual focus of social construction of technology or SCOT is that technological artifacts are underdetermined by the natural world, and social factors and the interests of the actors determine the innovation process. Technological artifacts are explained and their construction interpreted with respect to the culture of the community of technology practitioners. Finally, in interpreting technological change both failed and successful innovations should be analyzed symmetrically using the same explanatory devices.

In the Green Revolution we saw that artifacts and practices from a variety of fields such as biology, agronomy, and engineering coalesced within a framework of different economic, political, and cultural factors. HYVs of seeds, chemical fertilizers and pesticides, electric and diesel pumps, tractors, irrigation canals, roads, banks, extension services, agricultural universities, national and state governments, international research institutes, international donor agencies, and so forth are "socially constructed" in themselves, evolved under disparate contexts. However, when these entities were brought together a significant event took place. The SCOT model falls short when constructing the larger picture of how the traditional technological system of subsistence agriculture gave way to the modern market-based system. However, SCOT may be able to explain some features of the development of individual components like HYVs of seeds. But this is the main methodological thrust of the SCOT program. Therefore, "[I]n any attempt to understand technological development, it is axiomatic that the technological artifacts (broadly construed to include materials and processes) play a prominent part in the analysis."[4] Therefore, the central argument of the SCOT program, whether the "mild" or "radical" version, that the Green Revolution can be understood by analyzing a variety of social, political, cultural, environmental, and economic issues surrounding the various artifacts is the vindication of the relevance of this model. In this sense, the technological change in the Green Revolution was a "seamless web" of society, politics, economics, material artifacts, and institutions.[5]

By concentrating exclusively on micro-level events and artifacts, the SCOT model may miss the larger picture of the macro-level events in the Green Revolution. It would be difficult to account for the processes of technology transfer and institutionalization of technological knowledge in the form of research capacity. The assertion that technology is entirely socially constructed and that once a technology is selected, the social factors get "frozen" in the artifact, becomes problematic because of the tautological nature of the assertion. By asserting that only social factors can explain the technological change in the Green Revolution would end up in abstracting all the other extant factors to a determinist theory of social necessity dictating technological change, leaving aside any sense of contingency in the explanatory mode.

Actor-Network-Heterogeneous-Engineering Models

In terms of its explanatory power, the actor-network and heterogeneous-engineering models are more valuable than SCOT.[6] The basic methodological premise of both models is to build a system in which the elements, both animate and inanimate, are linked together to bring "closure" and foster "stability." In terms of explaining technological change, John Law contends that "no change in vocabulary is necessary; from the standpoint of the network those elements that are human or social do not necessarily differ in kind from those that are natural or technological."[7] While Law's protagonists are heterogeneous engineers, engineer-sociologists are the main agents of change in Callon's model. Both Law and Callon adhere to the symmetry concept for explaining social, cultural, technical, and other extant factors by the same vocabulary.

The network builders might identify the following as the elements of the Green Revolution network: peasant-farmers, wooden plows, traditional seeds, bullock carts, dwarf and semidwarf varieties of rice and wheat, modern high-yielding varieties of seeds, chemical fertilizers and pesticides, the government of India and its various departments, agricultural universities and research centers, international agricultural research centers, bilateral and multilateral donor agencies, researchers and extension agents, plant genes, irrigation canals, electrons, roads, markets, and so on. The method of network building is to construct and reconstruct the network by adding or eliminating elements so that "collisions" and "closures" can be understood, eliminated, or achieved in eliciting the problem-solving function of the protagonists. Contingency plays a key role in system building and concomitant technological change.

Because of its overdependence on contingency as a key explanatory variable for explaining technological change, it may appear that the Green Revolution occurred as a serendipitous event rather than an event that was caused by the implementation of actions based on planning. Although the exact nature of the outcome of technological change could not be predicted in advance, we have seen already that technological change can be constructed by purposeful actions. The case study amply shows that the technological change in Indian agriculture did not take place by chance. The need for change in the technology was clearly felt, not only by the protagonists who were affected by the problem, but also by those who came forward to help and from those whose help was solicited. These groups combined their efforts to eliminate the problem. The bottleneck was the low yield due to marginally declining land productivity and to poor seeds and lack of new technologies.

However, the technological change did not occur simply by stringing together all the elements (actors) mentioned earlier. In the actor-network model what we have is only a cluster of actants with no overall structure or order as to how they were brought together. It is not apparent how, where, or when a new actant is to be included or excluded. Lacking a heuristic with which to build the actor-network, these two models lack analytic rigor. Although the model is intended to breakdown the barrier between micro- and macro-level concepts and events, simply linking them together does not achieve that goal. Complex and dynamic processes like the development of research capacity, the transfer of knowledge, the development of high-yielding varieties of seeds, and such other factors cannot be represented by a single element (actant) in the network.

NEO-SCHUMPETERIAN EVOLUTIONARY MODELS

Technological Paradigm and Trajectory

According to Giovanni Dosi, technological change is characterized by paradigm shifts as is scientific change.[8] The evolutionary character of technological change is hinged on the assumption that the "selection"[9] of a particular paradigm is based on a large number of possible directions of development according to the momentum generated by the technological trajectory.[10] Dosi qualifies this assumption by resorting to the operationalism called "notional possibilities."[11] Some of these selection criteria included in the notional possibilities are marketability, profit potential, potential for automation, and corporate strategies to capture more market share, and so on. Technological progress and change is indicated by the continuous changes in technological innovation along the locus of a technological trajectory that is defined by a technological paradigm. According to Dosi, a technological paradigm is a "set of procedures, a definition of the 'relevant' problems and of the specific knowledge related to their solution."[12] The "exclusion effects" of technological paradigms constrain the practitioners of the technology to focus only on a limited range of technological possibilities (trajectories). Dosi views technological trajectory as the "direction of advance within a technological paradigm."[13] Technological trajectory is the normal problem-solving activity determined by the technological paradigm. Finally, a technological paradigm is selected based on several extant factors, which most notably are economic and institutional. The crucial factor in applying the Dosian model to the process of technological change in the Green Revolution is to find the appropriate technological trajectory and the technological paradigm.[14]

Following Dosi, the problem solution (paradigm) of the traditional agricultural system is constrained by the stipulations of traditional customs, social relations, geographic locations, natural forces, and preindustrial technologies. The technological trajectory in traditional agriculture may be identified as the practice of an agriculture dependent on complete submission to the dictates of environmental factors.[15] Lack of significant control over natural forces such as water, nutrients, weeds, and seeds severely limited the productivity of the land in traditional agriculture. Furthermore, in traditional agricultural communities, it is normal for technological innovation to have reached a plateau because of minimal outside support for economic development. In traditional agricultural communities, new inventions and innovations are rarely contemplated because risk taking is avoided rather seriously because of "rational expectations." The technological paradigm in traditional agriculture may be simply put as subsistence farming. The social, economic, class, and caste structures are dictated by the subsistence farming worldview. Having learned from previous experiences, peasant-farmers shun innovations because they try to avoid any more uncertain variables in their production function, such as using untested seeds or irrigation methods. The inability to manipulate and gain control over natural forces was understood to be the most important "bottleneck" in improving traditional agriculture. The powerful "exclusion effects" of subsistence farming constrained all attempts at modernization of the existing production relations and processes. A new outlook on agricultural practices, spearheaded by a new technology, was the most viable way to break out of the stasis in agricultural technology. The emergence of a "new worldview" in Indian agriculture paved the path for a new technological trajectory dictated by the new technological paradigm.

The new technological paradigm may be characterized as one in which agriculture is practised with the aim of producing for a market.[16] The technological trajectory may be construed as the increasing ability on the part of farmers, agricultural researchers, and technologists to manipulate and control natural forces. The prime achievement in this category would be the development of high-yielding varieties of seeds. Other important developments included chemical fertilizers, modern irrigation systems, new agricultural equipment (tractors, pumps, threshers, weeders, and so forth), and storage and transportation systems. The necessity to increase the productivity of the land, whose area was a fixed variable, in turn, determined the nature of the new technologies. The selection criteria may be identified as the political drive to attain self-sufficiency in cereals to attenuate dependency on foreign food and to alleviate rural poverty. Potential goodwill on the part of international donor agencies to help poor nations by transferring agricultural technology may be another criterion. This, indeed, could be to keep a large nation like India adhering to the "correct" economic and political path during the heyday of the Cold War.

According to Dosi, the development of the HYVs of seeds, for example, would be the outcome of a "mutation," that is, an *ex post* selection process similar to a Schumpeterian trial-and-error process. On this view, the HYVs became only a chance development based on the particular technological paradigm. Although contingency may have had an influence on aspects of their development, HYVs were not exactly a chance invention. Although the precise nature of a new invention cannot be predicted in advance (if we knew it in advance, it would not be an invention), the avenue through which the new inventive activity is being conducted and the purpose of the action are known *ex ante*. Similarly, in agriculture, farmers knew all along that one of the crucial factors in achieving better land productivity was finding the best seeds. Thus, the decision to establish research institutions that were intended for adaptive research to develop new seeds and the eventual development of the HYVs was not a chance occurrence. The Schumpeterian trial-and-error process, in which entrepreneurs come up with new technologies, may not quite closely follow in the case of the Green Revolution, as Dosi's model would postulate.

Technological Guidepost and Innovation Avenue

Devendra Sahal presents a metaevolutionary model of technological change and progress with several features similar to Dosi's model (it should be recalled that Sahal published his work before Dosi did). Sahal conceives of technology in a systems framework, in which the process of technological change is regarded in terms of certain "bottleneck" factors such as natural resources and efficiency considerations.[17] Sahal views technology in terms of functional and performance characteristics, and contends that technological change is a continuous process in which several minor changes cumulate over a period of time to bring about the major change in the technology.[18]

The processes of innovation and technological change, according to Sahal, are influenced and regulated by the boundaries of the technological system. For him, the dynamics of the system formation is the most important factor that charts the course of long-term technological change. For example, the emergence of a particular machine design is unlikely to be the result of the instantaneous creative genius of an engineer or a group of engineers. Instead, it may be the cumulative synthesis of the design ideas that were available in the field for a long period of time. The innovations that take place in the technology in question, at any particular moment, are only piecemeal modifications of a design that remains "unchanged in its essential aspects over extended period of time."[19] This basic design may be characterized as the "guidepost" that charts the course of technological change. In Sahal's schema, technological change does not happen in a "haphazard fashion," but instead it occurs in an "expected manner on what may be called *innovation avenues* that designate various distinct pathways of evolution."[20]

Technological change in the Green Revolution, according to the Sahal model, would be an "invariant design" acting as the technological "guidepost" that guides technological inventions and innovations. Even if we accept the invariant design concept in a metaphorical sense, it is very difficult to find a similar applicable condition in the Green Revolution. A major problem in identifying this concept in a macro event like the Green Revolution is that Sahal developed it using micro case studies of already embodied technology. Although the change of agricultural technology from traditional to modern can be characterized in evolutionary terms, there is no technological guidepost that is isomorphous in both phases. As we have already seen, the dynamics of growth and change in these two phases are different.

The technological guidepost of Sahal has several things in common with Dosi's concept of technological paradigm, and one may conflate both without losing conceptual rigor. While technological trajectories designate the loci of technological change in Dosi, innovation avenues designate the pathways of technological change in Sahal. Nevertheless, Dosi's model is more evocative and analytically superior than Sahal's. Dosi provides a clear heuristic as to how to locate the technological paradigm. Instead of offering a heuristic, Sahal only makes the claim that the presence of an invariant design acts as the technological guidepost.

ECONOMICS MODELS

Production Function Model

The dominant paradigm in the economics discipline is neoclassical economics, which postulates that technological change can be represented by a production function. A production function is a relationship between various combinations of inputs or factors of production and outputs.[21] The production function of a firm can be represented graphically as the plot of the isoquant, which is a representation of different techniques employed in producing the same output in which the axes are represented by the input variables. Technological progress and change is represented by an inward shift of the production function isoquant where the same output can be produced with fewer amounts of the inputs.

The production function of an industry or economic sector is the combined (aggregate) production functions of all the firms or units of production in the industry or economic sector. Robert Solow claimed that it is possible to have a single aggregate production function for a whole economy that is similar in nature to the production function of a firm.[22] Neoclassical economists claim that technological change can be ascertained by a residual method. They do this by first accounting for the increase in productivity due to their factors of production, and then assessing the residual (technological change) by deducting the change in productivity due to factors of production (capital and labor) from the overall productivity of the industry or the economy.

In order to apply the neoclassical production function model to explain a case study, the first task is to identify the unit of analysis, which is the "firm." In the Green Revolution case, the closest entity that can be identified as the unit of analysis would be the peasant-farmer households, both small- and medium-scale farmers who have been practising agriculture until the Green Revolution began.[23] The second major assumption would be the existence of perfect knowledge about alternative ways of production using new technologies. The third major assumption would be that there was a demand for the new technology and a ready supply, all under the assumption of elastic market prices and perfect competition.[24] Under these conditions, the households would engage in agricultural production by using inputs according to their relative prices, that is, they would engage in factor substitution.[25] The resultant technological change would be the forward shift of the isoquant. Technological change in the Green Revolution could be assumed as the progressive shift of the isoquant that represents the continued increase in productivity of the land.

The above characterization of the Green Revolution employing the neoclassical model is highly unrealistic and ahistorical. The problem with neoclassical analysis, as Jon Elster points out, is that it "is a supremely efficient tool for equilibrium analysis of economic life, including intertemporal equilibria, steady-state growth, and other phenomena that take place in *logical* as opposed to *historical* time" (emphasis added).[26] The economic and technological selection problems involved in the Green Revolution are dynamic, and the neoclassical model fails to account for the conceptual issues inherent in dynamic analysis. Serious difficulties arise in analyzing the technological change in the Green Revolution using the neoclassical model. In the case of the Indian peasant-farmers, almost all preconditions for engaging in factor substituting optimizing behavior were absent. There was no market (let alone a perfect market) in most rural areas before the onset of the Green Revolution.

Many prominent analysts of development challenged the adequacy and relevancy of neoclassical paradigm as a means to understand the process of economic development and technology transfer in the Third World. Gunnar Myrdal seriously challenged the stable equilibrium, comparative advantage, and free trade theories of neoclassical economics as inapplicable to the Third World.[27] Paul Streeten argued that concepts that are taken for granted in neoclassical economics such as capital, income, employment, price level, savings, and investment are absent in most Third World countries.[28] The India of the 1950s and 1960s did not have the conditions to apply the neoclassical production model of technological change.

India was a feudal society (and still is in many parts of the country) where 80 percent of the population depended on land, one way or the other. There was no market for credit (other than usurious moneylenders) for buying seeds or other inputs, and there was no market access for selling of the surplus produce. Most of the peasants were illiterate and hence the assumption of perfect knowledge of alternative production techniques was unrealistic (for that matter even in a society with perfect knowledge). The communication facilities to disseminate information in rural areas were minimal. In addition to these objections, the neoclassical model of technological change does not consider time as an important variable. The neoclassical model assumes that the production function at a particular instant reflects all the conceivable combinations (technologies) of production, including those available in foreign markets. If a factor of production available in a foreign market has a comparative advantage over the local factor, then the local producers would import that factor of production. This sort of an explanation for the process of technology transfer involved in the Green Revolution is unsatisfactory, if not deficient in historical justification.

Another serious problem with the production function model is the assumption that the peasant-farmers had several techniques from which to choose, and they chose the Green Revolution package of technology because of its relative price in comparison to other alternatives. This assumption holds that they were aware of the alternative courses of actions. We know that the peasant-farmers did not accept the technology the moment they heard about it. It took enormous prodding from the government. They accepted it only after they were convinced of the benefits of the technology through trial and error.

Induced Innovation Model

The induced innovation model that considers technological change as an inevitable by-product of institutional innovations may be able to explain the Green Revolution more effectively than its production function counterpart within the neoclassical tradition. As we saw earlier, the major thrust of the induced innovation model is that institutional change precedes technological change, and that institutional innovations occur when it becomes profitable for individuals or groups of individuals to bring about the change.[29] The induced innovation model can explain the technology transfer component of technological change much more effectively than alternative economic models.

According to the induced innovation model, technological change in the Green Revolution may be characterized as the result of the changes in the institutional structure demanded by the peasant-farmers. The institutions that experienced change, according to Ruttan, were the markets, peasant households, and public agencies.[30] However, the institutionalists revert to the less-beneficial production function model when they actually explain the process of change at the micro level. Ruttan and Binswanger argue that the changes in biological technology that led to the diffusion of the high-yielding varieties (HYVs) of seeds were the results of the "changes in relative resource endowments and factor prices."[31]

As Grabowski correctly pointed out, the institutions that the diffusion process changed were secondary ones such as markets and bureaucracies and not primary ones such as social relations, political power structure, ideology, and cultural mores.[32] The Green Revolution did bring about changes in some primary institutional structures because of changes in the means of production, and not the other way around, as the induced innovation model would claim. The induced innovation model made the ahistorical claim that "demand" for institutional change would "supply" technological change.[33]

The induced innovation theorists tried to remedy the lacunae in the orthodox neoclassical approach to explaining technological change by arguing that institutions are endogenous to economics, and that technological change in turn is endogenous to institutions. Nevertheless, by retaining such concepts of the neoclassical model as the production function,[34] optimization, and maximization, the induced innovationists seriously undermined their efforts. The induced innovation model's ability to explain the process of adoption of a new technology may be further aggravated when contrasted with the methodology of technological innovation developed in sociology and anthropology.[35]

As indicated earlier, the induced innovation model can explain the technology transfer component of the technological change in the Green Revolution with greater facility. Induced innovationists delineate three "phases" of international technology transfer: (1) material transfer, (2) design transfer, and (3) capacity transfer.[36] Hayami and Ruttan argue that the Green Revolution falls under the third category, whereby there was a massive transfer of "scientific knowledge and capacity." This aspect of the technological change, in the case of the Green Revolution, can be corroborated with reference to the case of India. During the Green Revolution, in India there was large-scale creation of research capacity for the production of locally adapted technology. The development of a highly successful, indigenous agricultural research capacity in India was a capacity transfer of somewhat similar technological knowledge that existed in the United States during the 1940s and 1950s. However, large-scale transfers of materials such as HYVs of rice and wheat seeds and the technological expertise to construct and operate agricultural industrial ventures were undertaken as part of the technology transfer programs.

CONCLUSION: DO THE MODELS CONVERGE?

The neoclassical production function model accounts for some of the artifactual changes in the Green Revolution. It would explain the change from traditional to modern forms of agriculture as a random process of choosing the best technological practice from a "technology shelf" by the economic agent (peasant-farmer). In effect, the model explains the impact of the change, because what it explains is quantitative measures of change in productivity, rather than the change in such profound qualitative factors as the change in the knowledge related to the practice of agriculture itself. For all practical purposes, the model relegates technological change to a black box.

The induced innovation model, although explicitly conceding that technological change is endogenous to institutional change, contends that individuals and communities opt for technological change in a roundabout way, by first opting for a profitable institutional change. That is, the new institutions act as determinants of a new set of technologies. According to this view, the peasant-farmers of India opted for the Green Revolution because they sought change in their institutional structures. Although this model is inherently superior to its orthodox neoclassical counterpart, the creators of this model analyze changes only in secondary institutions like markets and bureaucracy, instead of changes in such primary institutions as social relations of production, structures of communication, and political power distribution between different social groups.

A. P. Usher provides a four-step heuristic to explain the cumulative synthesis of technological change using the unit of analysis of artifact. This heuristic can explain the invention of individual elements of the Green Revolution network. However, in a model that emphasizes problem solving as its central explanatory feature, the aim should be to make this process intelligible and hence the unit of analysis should transcend the artifacts to the human agents. In the Green Revolution, the agents (subjects) are peasant-farmers, and the process of changing their knowledge base is the central feature of the technological change. The cumulative synthesis model does not explain the learning processes and the institutionalization of a research system in the Green Revolution. The goal of the problem-solving activity in Usher's model is external adaptation of artifacts to follow a processual sequence, instead of an internally consistent act of modifying the knowledge base of the subjects and their sociotechnological system.

The unit of analysis of Hughes's model of technological change is a system. The development of a system, through which technological change can be analyzed, is characterized by a sequence of five phases. Although a few of the five phases can be identified for the Green Revolution, delineating the boundary conditions or parametric values for the Green Revolution is rather problematic. The major focusing device for the reverse salient in the Green Revolution was the low productivity of land and the lack of new knowledge on high-yielding seeds. Hughes's model, having been developed for industrialized nations, does not fit well as an analytic tool for preindustrial and precapitalist societies. Having not presented a heuristic of how the knowledge change of the agents actually happens, the model has limited applicability to explain the technological change in the Green Revolution.

In the social construction of technology or SCOT framework, the construction of artifacts, and by default the process of technological change, is explained with respect to the culture of the community within which the development of the artifact takes place. The "closure" of problems or the solution of technological "puzzles" is achieved by finding appropriate solutions based on the "interests" of the technology-developing communities. SCOT would explain the development of individual artifacts in the Green Revolution, such as HYVs, fertilizers, pesticides, irrigation pumps, and so forth, but would fail to explain technological change in itself as the event. The model cannot explain how the different structural components of technological change are constructed. It conceives the problem-solving nature of technological change as an external event that was induced to transform the artifacts, instead of modifying and changing the knowledge base of agriculture practised by the peasant-farmers, as was the case in the Green Revolution. The reason for this important omission was due to the model's assumption that once a particular technology is chosen and constructed, it takes care of the internal issues concerning the subjects. The model further assumes that social factors are frozen in a given technology, and that technological change occurs independently of social factors. This determinist trap can be avoided only by a reversal of its unit of analysis from artifacts (objects) to the actors (subjects).

Like the social constructivist model, the actor-network model emphasizes that there should be no division between the social and natural worlds. An actor network is constructed by stringing together all the tangible and intangible elements surrounding the technology in question. These elements are joined together in such a way as to eliminate problems ("closure") and create stability. Contingency is the key explanatory variable for explaining technological change. Although contingent factors did play active roles, the technological change in the Green Revolution was the result of a carefully planned and executed program. The actor network is an excellent explanatory device at the micro level, but it fails to explain events that transcend the immediate microworlds of the actors.

In terms of interpreting the determinants and explaining the directions and dynamics of technological change in the Green Revolution, the Dosian model is perhaps the most appropriate of all the models analyzed in this chapter. Its explanation of contextual factors in the selection of technological paradigms is cogent. In spite of this, there remain some serious problems. Dosi's primary focus has been on technological change in market-oriented capitalist economies. Most of the determinant factors that Dosi presents can be found in such systems and hence its applicability to the economic conditions of rural India poses problems. Furthermore, Dosi's adaptation of the Kuhnian paradigm from the context of scientific change to technological change is amenable to the same ambiguities that confound Kuhn's paradigm. Dosi's claim that he finds considerable epistemological congruence between scientific and technological paradigm may be controversial given the epistemic differences between these two enterprises, as explained in chapter one.

Although no consistent pattern of explanation emerged that allowed us to explain the important stages and events, each of the models, with a few exceptions, was able to account for some of the salient features of the Green Revolution. It was also observed that major differences persist, according to disciplinary perspectives and prerogatives, in accounting for what is considered important facets of technological change. These models apparently were developed to explain and account for the same phenomenon, viz., *technological change*. The lack of their convergence in terms of their ability to account for the same phenomenon (encapsulated in a case study) is an area that requires scrutiny. This will be addressed in chapter nine. It should be mentioned, however, that the Green Revolution is only one case study. More authoritative accounts of convergence can only be ascertained by testing the models against other comprehensive cases of technological change. However, testing against multiple models does not guarantee that convergence would occur. Such an exercise may only vitiate the problem further. Those interested in such investigation may follow the lead provided here and test convergence using other case studies.

ENDNOTES

1. The particular "inventive" steps that created the semidwarf varieties are difficult to delineate from the historical materials available. One can only suggest that continuous crossbreeding between plants with dwarfing characteristics might have created the semidwarf varieties. Conversely, they may have been the result of evolutionary processes involving random mutation.

2. It may be recalled that Borlaug was awarded the 1970 Nobel Peace Prize for his pioneering works in plant genetics, which supposedly alleviated the problem of world hunger.

3. Abbot P. Usher, "Technical Change and Capital Formation," in *The Economics of Technological Change: Selected Readings,* Nathan Rosenberg, ed. (Harmondsworth, UK: Penguin Books, 1971), 43–72, 50.

4. Trevor Pinch, "The Social Construction of Technology: A Review," in *Technological Change: Methods and Themes in the History of Technology,* Robert Fox, ed. (Amsterdam: Harwood Academic Publishers, 1996), 17–35, 23.

5. Thomas P. Hughes, "The Seamless Web: Technology, Science, Etcetera, Etcetera," *Social Studies of Science* 16 (1986): 281–292.

6. Since the actor-network and heterogeneous-engineering models are conceptually and methodologically closer, we will conflate both to explain the Green Revolution.

7. John Law, "Technology and Heterogeneous Engineering: The Case of Portuguese Expansion," in *The Social Construction of Technological Systems: New Directions in the Sociology and History of Technology,* Wiebe E. Bijker, Thomas P. Hughes, and Trevor Pinch, eds. (Cambridge, MA: MIT Press, 1987), 111–134, 114.

8. Giovanni Dosi, "Technological Paradigms and Technological Trajectories," *Research Policy* 11, no. 3 (1982): 147–162. Dosi posits a close epistemological parallel between science and technology. Although Dosi adds a caveat that this parallel should be construed in an impressionistic sense, he, nevertheless, moves beyond his own self-imposed restriction, and almost wholeheartedly adopts a pipeline model of science-technology relationship that was rejected by George Wise, "Science and Technology," *OSIRIS,* 2nd Series, 1 (1985): 229–246, and others as historically inaccurate.

9. According to Dosi, "Technological Paradigms," 153, the "selection" process happens as follows. First, Dosi considers the "downward" sequence of "science-technology-production" with all the limitations considered of such an analysis. Beyond pure science, many scientific theories and puzzles are passed on to the realm of "applied science" and technology (the boundaries between these two is not clear). Thus, in the science-technology-production scheme the economic forces together with the institutional factors operate as the selection mechanism.

10. Because of this "directedness" of the selection process, Dosi's model diverges from the randomness associated with natural selection in biological evolution.

11. As will be shown in the case of the Green Revolution, these "notional possibilities" may take the form of competing technological paradigms, thus blunting the very possibility of a reigning paradigm in the historical evolution of a technology.

12. Dosi, "Technological Paradigms," 148. It is more like a "worldview" in the Kuhnian sense.

13. Dosi, "Technological Paradigms," 148.

14. Although the Kuhnian and Lakatosian models of scientific change are primarily intended for comparative judgement of competing paradigms or research programs, Dosi

tends to apply these concepts for cases of technological progress in isolation. That is, he does not offer empirical support to argue how one not so successful technological paradigm replaces a successful one. For example, Dosi does not go into the details of technological change in the electronics industry while it was in the vacuum tube mode. Instead, he discusses the various progressive problem-shifts in the semiconductor mode.

15. Of all human endeavors, agriculture is most heavily dependent on environmental factors. The progressively efficient technologies with which humans were able to control environmental factors were direct variables in increasing agricultural productivity in any social system.

16. Besides the increased monetization and commodification of the rural economy, producing for a market involves a much more comprehensive change, the most important one being the transformation of peasants into farmers.

17. Devendra Sahal, "Alternative Conceptions of Technology," *Research Policy* 10 (1981): 2–24, 13.

18. This appears very similar to the theory of technological change postulated by A. P. Usher (the theory of cumulative synthesis).

19. Devendra Sahal, *The Patterns of Technological Innovation* (Reading, MA: Addison-Wesley Publishing, 1981), 33.

20. Devendra Sahal, "Technological Guideposts and Innovation Avenues," *Research Policy* 14 (1985): 61–82, 71, emphasis in original.

21. The factors of production are labor and capital, and the outputs are corn, guns, butter, and so on. Note that unlike their classical predecessors, neoclassical economists do not include natural capital in their input function. They contend that natural capital is "unlimited" and it has no "marginal value."

22. Robert Solow, "Technical Change and the Aggregate Production Function," *Review of Economics and Statistics* 39 (1957): 312–320.

23. Assuming the households as "firms" begs many questions. Unlike the firms in an industry, peasant households are a "natural" entity most of whom do not become farmers by choice, but by tradition. The households in capitalist economic systems may be approximated to the firm, particularly in the United States and other industrialized nations, where large-scale commercial farming dominates the agricultural sector. But in a semifeudal and precapitalist economic system like in India, the peasant households can be hardly equated to the firm in the neoclassical framework.

24. The prices of both the technology and the agricultural products have to be assumed elastic.

25. A simple case scenario would be as follows: If the relative price of labor was high, then the households would substitute labor with labor-saving technologies. On the other hand, if the price of capital was higher than labor, the households would substitute capital with more labor.

26. Jon Elster, *Explaining Technical Change: A Case Study in the Philosophy of Science* (Cambridge: Cambridge University Press, and Oslo: Universitetsforlaget, 1983), 100.

27. Gunnar Myrdal, *Economic Theory and Underdeveloped Regions* (Bombay: Vora & Co., 1958).

28. Paul Streeten, "The Use and Abuse of Models in Development Planning," in *Leading Issues in Economic Development,* Gerald M. Meier, ed. (New York: Oxford University Press, 1976), 875–889.

29. Vernon W. Ruttan, "Induced Institutional Change," in *Induced Innovation:*

Technology, Institutions, and Development, Hans Binswanger and Vernon Ruttan, eds. (Baltimore: Johns Hopkins, 1978), 327–357; and Richard Grabowski, "The Theory of Induced Institutional Innovation: A Critique," *World Development* 16, no. 3 (1988): 358–394.

30. Ruttan, "Institutional Change."

31. Vernon W. Ruttan and Hans Binswanger, "Induced Innovation and the Green Revolution," in *Induced Innovation,* Binswanger and Ruttan, ed., 358–408.

32. Grabowski, "Institutional Innovation." The institutionalists do claim that theoretically their model applies to primary institutions. But to argue that changes in these fundamental institutions can be created by price incentives is unsupportable.

33. Cognizant of this undefendable position, Ruttan amended his position to argue that technological change and institutional change follows a dialectical relationship. Change in one induces change in the other, instead of a one-way relationship of change in institutions inducing change in technology.

34. The production function that the induced innovationists retain is a "metaproduction function," which includes land, labor, physical capital, human capital, and R & D. This certainly is an improvement on the orthodox neoclassical version where only two factors, capital and labor, are included.

35. Everett Rogers, *Diffusion of Innovations* (New York: Free Press, 1983), 11, defines innovation as the process by which: (1) an innovation (2) is communicated through certain channels (3) over time (4) among the members of a social system.

36. Vernon W. Ruttan and Yujiro Hayami, "Technology Transfer and Agricultural Development," *Technology and Culture* 14, no. 2 (1973): 119–151; and Vernon Ruttan, "Technical and Institutional Transfer in Agricultural Development," *Research Policy* 4 (1975): 350–378.

9

Technological Change as Knowledge Change

Toward a Syncretic Theory of Technological Change

THE EPISTEMIC SIGNIFICANCE OF TECHNOLOGY

The intellectual autonomy of technology as a knowledge enterprise has been well established. It was shown in the first chapter that the prevalent assumption of treating technology as applied science is a nonstarter in any attempt to understand the process of technological change, or for that matter, to understand the process of scientific change. An intellectual awakening in Science and Technology Studies (S&TS) has clearly established the autonomy of technology from science, and vice versa. The relationship between technology and science is one of mutually beneficial interdependence. Science and technology are intellectually and epistemologically autonomous knowledge enterprises, but related to each other dialectically. Although technology is autonomous with respect to science and vice versa, neither is autonomous with respect to society.

It must be emphasized, however, that the contextual factor that was most responsible for establishing the significance of technology as a knowledge system was the attribution of an interactive relationship between technology and science.[1] Although attributing a knowledge component to technology is as old as the discipline of the history of technology itself, this view solidified into a specific analytical category only after the acceptance of the technology-science interactive relationship as a model to understand the social and epistemic bases of these human activities. In this model not only are technology and science seen as autonomous epistemic enterprises, but also the knowledge component associated with these enterprises is not assumed to be uniquely privileged. That is, technology and science are not autonomous with respect to the larger domains—society and culture—in which they are only subsets.

171

A major outcome of the acceptance of technology as an autonomous enterprise has been the recognition of "technology as knowledge."[2] Technology should be perceived as a body of knowledge with its own internal dynamics of change and progress. Ideas, information, and other manifestations of knowledge rather than material artifacts are the primary building blocks of technology.[3] Technology has technological laws (which might often be tacit in nature), which help to create new designs, new products, and new processes.[4] These laws may be codified in the form of tacit design methods and styles of practice of the engineering and technology professions. Technological knowledge is tacit to the extent that its transmission is possible without being presented as certified knowledge, as is the case with scientific knowledge. This transmission is possible through craft traditions, family practices, and shop floor apprenticeship programs. However, in most advanced industrial societies, this form of technological knowledge production and transmission is less prominent as large segments of technological knowledge have become certified knowledge that is taught in engineering and technological schools, and have become part of the larger social domain.

The epistemic significance of technological knowledge stems from its clear dependence and interfacing with society and culture. Technological knowledge is produced due to the change in the means of production of the dominant social groups. It is not an anonymous accumulation of merely material artifacts, but the growth of knowledge related to the social production and other related activities conducted by individuals and social groups for solving their problems. Thus, technological knowledge production is first and foremost a social and cultural activity that is undertaken within particular political and cultural contexts.

The fundamental building blocks of technology are ideas, and technology is both a public and a private "good."[5] Compared to other forms of knowledge, technological knowledge is more easily kept private and secret through intellectual property rights regimes. However, at the same time, the artifacts and processes generated from such knowledge have to return to the public realm of economic systems, mediated largely through the market mechanism for their dissemination, diffusion, and ultimate survival. This paradoxical situation is unique to technological knowledge. Scientific knowledge, on the other hand, has to remain in the public domain in order to be accepted as certified knowledge.[6]

Ideas form the core of technology. In effect, technology is comprised of mental and physical activities that are based on ideas and thoughts of human engagement. The idea to use a stone or a stick for extending the reach and learning potential of the human intellect evolved as a cognitive process.[7] The externalization and ultimate materialization of the cognitive responses to the natural and social worlds was borne out of ideas, thoughts, and ultimately of human experiences. This was first accomplished with the simple tools and eventually with the sophisticated machines, instruments, computers, and other artifacts of modern times. The development of technology as a knowledge system is attributable to the increasing sophistication of human cognitive activities and processes.

THE CONSTRUCTION OF TECHNOLOGICAL CHANGE

Technological change is both socially constructed and political-economic in nature, because the agents and actors who make the decisions and policies regarding the development and selection of new technologies are particular social groups, who are also motivated and influenced by politics, economics, institutional imperatives, and cultures in which they are players and makers of the script at the same time. The development of technological artifacts and the whole process of technological change are embedded within societal, economic, and political considerations.[8] In the social construction of technological change, technological artifacts, society, politics, and economics become a "seamless web" of actors that are brought together to solve particular human needs and problems.[9] At a superficial level, being members of societies, all of our actions, beliefs, knowledge, and policies are social constructions, and consequently technological change also does not fall outside this norm of constructed construction. It is not the apparent tautological nature of technological change under the social constructivist framework that is stressed here, but the difficulty of constructing models and theories of technological change in order to come to terms (understand) with a socially constructed activity. Hence, analytical models and explanatory approaches to understanding technological change should not jettison the political-economic arguments and considerations, including the dynamic interplay of power, that shape and influence technological development. In fact, the very presence of power and political-economic considerations behind this process should animate theorizing and model building.[10]

What influences the direction and shape of technological change are environmental and resource constraints, governmental regulations, national and local politics, class and group alignments, taxation policies, investment priorities of governments and venture capitalists, R & D support from public and private sources, and consumer preferences (to some extent). It must be conceded that these factors do not have an equally weighted influence on this phenomenon. The course and magnitude of technological change has taken a qualitatively different form ever since the government became an influential macro agent in formulating science and technology policies.[11]

Although the political, economic, and social interests of the dominant groups in society set the stage for technological change and determine its trajectory, the actual process of learning and adaptations required for that change are dictated by the cultural and cognitive attributes of the different members of the social groups involved. While the social, economic, political, and other extant factors set the stage for the change, the internal dynamics of the actual change are shaped by, to a large extent, the functional attributes of the technology. Some of these functional requirements are the potential for energy substitution, the possibilities for automation for avoiding hazardous situations and deskilling and controlling workers, the search for ways to increase worker productivity, the ability for invariant reproduction of products and services based on the same ideas and design concepts, and new methods of communication to sustain the newly constructed technological ordering of humans and nature.

Technological change is a continuous and to some extent a cumulative process. However, we do not yet have the ability to predict and anticipate the exact nature and future course of technological change. We are limited to the ability to point out specific trends and styles. Historical narratives of technological change clearly demonstrate the cumulative and evolutionary nature of this phenomenon. What is meant by cumulative is that ideas, practices, laws, and theories from the past pass on to the development of newer technologies.

Technological change is essentially an evolutionary process. It must be recognized at the very outset that technological change is structurally and practically quite different from biological evolution. "Evolution" is used here as an explanatory metaphor to highlight the temporal nature of technological change, and nothing more. As Chris de Bresson and Joel Mokyr correctly point out, technological evolution is not analogous or isomorphous to biological evolution.[12] A better analog might be an ortholinear or directional variation, rather than a random variation as one might construe by adopting the evolutionary metaphor. However, it was noted in chapter five that some analysts use it in more than its metaphorical use-value. Technological change, from historical hindsight, appears as a selective variation process that is adapted to a sequential process of variation and selection. The selection and retention processes are characterized by instruction, understanding, experience of learning by doing, and, finally, cognitive change. Within the larger context of social and cultural evolution, adaptive learning and perception lead to the accumulation and change of technological knowledge. The evolutionary metaphor is important in explaining technological change because unlike the models available in certain fields (for example the production function model in neoclassical economics), it captures the temporal dimension of this phenomenon and reinforces the significance of historical trends, hindsight, and contingency.

The evolutionary metaphor is also useful to highlight the often contingent nature of technological change. Although the policy instruments that are required to shape technological change to follow an ortholinear or directional variation may be able to set a particular trend, contingent factors can change or reshape this trend. Unexpected political, economic, and environmental factors may often make the long-term policy guidelines that will shape the course of technological change too costly and/or ineffective. A case in point is the failure of cereal production in India in the 1960s to feed the population, which in turn altered the course of the ambitious industrialization campaigns of the Nehru government that started in the 1950s. The decision to modernize agriculture, under pressure from aid donors and foreign powers, as a means to increase food production by inducing technological change through technology transfer and other mechanisms, was forced upon the government to address the contingencies created by the near-famine situation in the country. These measures altered the course of industrialization and the nature of technological change in India. The technological change that came about as a result of the Green Revolution analyzed in the previous chapters was a classic example of knowledge change shaped by contingencies.

The cumulative nature of technological knowledge explains the irreversibility of technological change. It is not meant to imply that past experiences from the historical evolution of the technology in question should accrue following a uniformitarian theory of accretion. By cumulative, what is meant is that the functional attributes and the basic principles of design and operation of technology, to a large extent, remain invariant. The fundamental functional attributes of a technology remain more or less unchanged, unlike in science where paradigmatic changes occur due to new discoveries or new interpretations of observations. The end use of all technological artifacts remain the same, such as a wooden plow or a mechanical tiller or tractor. In the case of the wooden plow, the work is done more slowly using human and animal power, while in the case of the tractor, the work is done by mechanical power (derived from fossil fuels) at a faster rate. The functional attributes of both remain invariant. The changes are confined to the modification and adaptation of existing designs, operating procedures, materials for constructing the new variations, and to the energy sources. It appears almost axiomatic that, fundamentally, technological change is shaped by its antecedent technologies. It is thus imperative that models and theories of technological change should account for the evolutionary, contingent, and cumulative nature of this phenomenon.

TECHNOLOGICAL CHANGE AS KNOWLEDGE CHANGE

The models of technological change analyzed in this book do have structural and operational similarities. They do explain some aspects of this phenomenon. However, beyond this, their explanations of technological change did not converge. An important reason for this failure is that most of these models did not transcend the artifactual realm of technology in their theoretical expositions of technological change, as changes in the constitution of the artifact do not necessarily capture the complete gamut of this phenomenon. Besides, what constitutes artifactual change can vary according to the differing perception of an artifact.[13] As a result, most models tend to focus on different structural components like invention, innovation, development, and transfer. In order to explain technological change, models should recognize the epistemic basis of technology and should consider technological change inherently as knowledge change.

It was shown in the previous chapter that the important models of technological change from history, sociology, economics, and evolutionary studies could not consistently explain the process of technological change in the Green Revolution. Although these models were well constructed and presented, they failed to converge on explicating the same phenomenon encapsulated by the Green Revolution. Taking these models as a first step in constructing such a complete model, the following Eleven Theses are presented to suggest what features and explanatory concerns a comprehensive model of technological change should have, using the Green Revolution as the empirical basis for explication wherever possible and necessary. This exercise takes the issue of reflexivity and self-referencing seriously. However, the objective is not to falsify any of the models discussed in this book, which is an impossible task in itself and an intellectually futile venture as well. On the contrary, it is undertaken in a constructive spirit of improving the existing models and to help those interested in developing a new model of technological change that will acknowledge the theoretical precept of technological change as knowledge change. However, ultimately, the objective is not only to represent technological change, but also to intervene by representing this social phenomenon in a meaningful way.

ELEVEN THESES ON TECHNOLOGICAL CHANGE

Technological change should be conceptualized under the premise of technology as knowledge

The question concerning technology is how to reveal its epistemic significance. The question concerning technological change is how to theorize and explain this dynamic and multidimensional process that is embedded within societal, political, economic, historical, and cultural factors. Technological change is essentially a contingently shaped social-historical process with agency. Technology should be seen as a body of knowledge and practices concerned about practical problems and solving problems related to human needs and wants. Technology is more than its material and physical constitution. Ideas and information and other manifestations of human knowledge characterize technology. Technological knowledge may often appear tacit and its transmission is possible without being presented as certified knowledge as in the case of scientific knowledge. It must be highlighted that the epistemic significance of technology as knowledge is due to its interfacing with society. Technological knowledge is produced essentially as the outcome of the changes in the means of production of the relevant social groups in society. Technological knowledge production is a social-cum-cultural activity undertaken within particular political and economic contexts.

It has been shown that the Green Revolution illustrates a rich historical account of a technological change. It is more than a chronicle of the invention, development, transfer, and diffusion of artifacts and techniques. This case study reveals that technology entails a sound knowledge of ways and means of doing things to solve a problem. The Green Revolution entailed a package of technology whose knowledge content evolved to circumvent the constraints of low productivity of the land. It cannot be decoupled from knowledge, because what essentially took place during the technological change was the change of an "old" knowledge system to a "new" knowledge system, such that the latter solved the problem much more effectively than the former.

The creation of this new technological system was achieved by transferring significant aspects of a knowledge system developed in the United States and elsewhere to India. This included the development of a research capacity to adapt that knowledge on seeds, seed preparation, nitrogen and fertilizer fixing, pest management, watering, and so on to the specific needs and conditions of the recipient country and its transmission, specifically to the peasant-farmers who ultimately learned the new system through trial and error, learning by doing, learning by using, and such other learning and information exchange processes.

The knowledge behind the creation of hybrid rice varieties was known to the Chinese thousands of years before the modern high-yielding varieties (HYVs) of seeds were developed. Crossbreeding techniques of developing short-strawed rice and wheat form the basis of this practice. The Japanese who took these varieties to Japan (from its colonies, Korea and Taiwan) and experimented with chemical fertilizers to increase productivity further improved these short-strawed varieties. United States scientists transferred some hybrid wheat varieties from Japan where the progenitors of the modern HYVs of wheat evolved. The development of the HYVs was made possible, in part, by the knowledge that already existed in the form of genetic manipulation of crossbreeding, a form of biotechnology. Once transferred to India, the HYVs underwent genetic manipulation to sustain their high-yielding potential in the context of different climatic conditions and culinary preferences. The creation of research and development facilities and institutions of higher learning made this action possible. Finally, the knowledge of the experts had to be translated to the peasant-farmers through an elaborate network of state and local level extension and outreach programs. Thus, the technological change involved a complex array of information transfer, creation, and diffusion among the actors.

Technological change should be conceptualized in systemic terms such that technology is defined as an instrumental ordering of human experiences with efficient action as the plausible selection criterion

The concept of system does not automatically entail vertically integrated technical ensembles and engineering production systems with strictly delineated boundaries. It is meant to denote a network of technological and social entities connected in a horizontal mode that can be circumscribed by certain boundary conditions to construct the sociotechnological system for effective problem solving. The goal of such a sociotechnological system is to create proper structures that are amenable to being controlled to achieve planned goals and policy objectives.

The Green Revolution reveals that the technological change evolved as the "traditional" agricultural system gave way to a new one with a different dynamic. In the old case, the boundary of the sociotechnological system did not stretch beyond the village. With the new one, it stretched far beyond the village. As a result of the flow of knowledge of agriculture through increased communication facilities, the sociotechnological system expanded to include markets, rural cooperatives, banks, roads, irrigation canals, extension networks, national and international agricultural research institutes, and state and federal governments, among other actors.

Both systems of agriculture were "efficient" when viewed from the narrow rational perspectives of their objectives. In terms of energy efficiency, the traditional system was more efficient than its modern counterpart. In terms of crop volume and productivity, the modern system was far more efficient than the traditional one. When land became a constant factor of production, the most efficient means to increase land productivity was by changing other variables in the production equation (process). Accordingly, the knowledge component became the most important variable to achieve this end.

The development of the HYVs shows that these new seed varieties came about as the outcome of a knowledge system that evolved out of rich human experiences dating back hundreds or even thousands of years. The basic idea behind these "miracle" seeds was energy efficiency. A smaller stalk meant fewer nutrients wasted in the stalk. However, its full potential began to emerge only in the late nineteenth and early twentieth centuries when chemical fertilizer was being used in agriculture. The new package of agricultural technology associated with the HYVs-fertilizer-irrigation system was clearly the result of an instrumental ordering of human experiences and ideas with efficient means as a fundamental principle of technological change.

Artifactual change is (only) the manifestation of the process of technological change

A complete model of technological change should specify that artifacts are the physical manifestation of the knowledge change at the primary level, while the knowledge change is manifested at the secondary level as social capital and knowledge, and at the tertiary level as specific cognitive traits of the individual agents. While technological change is an irreversible phenomenon, artifacts may be temporal entities. The reason why the former does not disappear while the latter does is due to the fact that the former involves something fundamentally different from the latter. This fact is directly related to the notion of "technology as knowledge," which is shown clearly in Thesis One above.

In the Green Revolution we saw that technological change was not characterized by the mere changes in the constitution of material artifacts such as seeds, pumps, fertilizers, plows, and so on, but the change in social relations, and ultimately the change in the knowledge base of agriculture practiced by the peasant-farmers.

**Invention, innovation, development, and transfer of technology
are (only) the component vectors of technological change,
and these act interactively and do not follow a lawlike sequence**

Technological change may involve all or some of these component vectors, but
not necessarily in that specific order. Technological change in the Green Revolution shows that these components were present in its construction. The HYV of
seed is an invention that was perfected through intensive collaborative work, especially through location- and context-specific research and development activities in several countries. It was transferred to India from the outside, specifically
from Mexico and the Philippines. The knowledge of building agricultural research
capacity through proper institution development was transferred largely from the
United States, which included the transfer of the know-how of training and visit
(extension work). Once the research capacity was created, new seed varieties were
developed in India. Feedback from peasant-farmers, extension agents, and consumers provided much of the background for the adaptive research. There was no
clear pattern of a lawlike sequence involving the various component forces.

A network of causal linkages in the Green Revolution can be demonstrated as:
invention (hybrid seeds, HYVs, HYV seed-fertilizer-irrigation agriculture)⇒*transfer* (of the new HYVs and the new agriculture practice to India; transfer of research
capacity)⇒*innovation* (further adaptive research in India and elsewhere to make
the HYVs location-specific)⇒*development* (establishment of research capability;
development of several new varieties)⇒*diffusion* (transmission of the new knowledge to the peasant-farmers through extension work) ⇒*feedback* (problems of new
varieties from peasants and consumers relayed to the researchers)⇒*development*
(further research and development work)⇒*transfer* (of the new knowledge of agricultural practice to other parts of India and other developing countries). This is an
open-ended network. Beyond certain predictable initial conditions, contingencies
shape the rest of the sequence. These entities or vectors acted interactively without
any hierarchical structure to bring about the technological change.

**Technological change is an evolutionary process operating within a
contingent social milieu**

Society and its institutions are not stable and unchanging entities, but rather dynamic entities that are affected by both internal and external sharing, tensions, and
conflicts. Although radically new ways of solving problems can emerge in the technological sphere, technological change is characterized by evolutionary change.
This is because what we understand as technological progress is a new problem-
solving culture that involves superior ways of doing things with radically new
ideas. The change is related to doing and solving the same old problem, a practical
matter concerning the agent.

The evolutionary nature of technological change in the Green Revolution shows that the evolutionary metaphor is not isomorphic to biological evolution. It is meant as a simple guidepost to show that technological change may be best understood as the result of a variation and selection process with efficient means as the criteria for selection within limited possibilities. The development of the HYV of seeds took a long period of time, beginning with the hybrid short-strawed varieties. The selection process of the HYVs developed in India followed a clear pattern of retention of the high-yielding potential with location specificity being a key independent variable.

Similarly, the process of technology transfer and the creation of agricultural research capacity in India involved decades of carefully carried out selection of those features of American agricultural education and research capacity building conducive to India, a newly independent country (that shed its yoke of colonialism only a decade ago), with vastly different economic and social conditions. The peasant-farmers learned the new agricultural practice through different learning processes. These processes were comprised of model demonstration programs, trial and error, formal and informal training programs, and training visit by the extension agents, among other steps. The change from the "old" to the "new" was not discontinuous. The decision to use the HYVs was gradual and was based on "rational expectations" of the peasant-farmers. Initially they experimented with the HYVs by devoting a small fraction of their land to the new seeds. They accepted the new seeds-fertilizer-irrigation technique only after being convinced of the high-yielding potential through a selective retention process of learning by doing.

In the Green Revolution, we also noticed that natural disasters, inadequate organizational structures, preindustrial technologies, demands for better living conditions, resource availability, and donor decisions influenced the variation and selection process. The joint actions of the various international donor agencies and the government of India created the selection environment (a form of metamarket). We saw the creation, transfer, and local adaptation of the new agricultural technology ultimately transform the existing system in some states where the Green Revolution made a lasting impact. The different stages in the technological change included understanding the new techniques and problems, organizing a new research system, and learning new ways of doing things by trial and error. The government created much of the early scientific and technological infrastructure, instead of a "free market," as part of this selection environment.

Technological change is a continuous and cumulative process

The idea that technological change is cumulative is closely linked to its evolutionary nature. By cumulative, it is not implied that all past experiences from the history of the particular technology is cumulated, and that technological change is merely an accretionary process. It is meant in the sense that ideas and functional attributes of the technology, to a large extent, remain invariant. Models of technological change ought to affirm the cumulative nature of this phenomenon because it is influenced by its antecedent technologies.

The driving force behind technological change is the limitation of the existing way of solving a problem using the available technological knowledge. In addition to the limitations of the existing tools, constraints imposed by the surroundings can also lead to the problem-solving effectiveness. Increased productivity, improved efficiency, less cost, fewer human interventions, and other such efficiency measures are normally achieved by improving on the existing tools and techniques. The Green Revolution shows that the knowledge change involved new ways of doing things; however, the fundamental nature of practicing agriculture did not change.

Technological change is a problem-solving activity

Technological change occurs as the outcome of a specific set of concrete actions and learning experiences of a wide array of actors. Analyses of technological change should demonstrate how the problems were defined and how solutions were reached through strategic planning by the agents. Although technological change is a problem-solving activity conducted in need-based or need-induced circumstances, the solutions of the specific problems identified during these circumstances do not come off a technology shelf as some models of technological change claim. Extant political, economic, social, and institutional factors, including government policies and firm strategies, organize the problem-solving activities. However, these factors by themselves do not provide the solutions. Technological problems vary from case to case and their solutions vary according to the degree of complexity of the technological system. Some of these problems are low efficiency, adverse environmental conditions, simple functional failures, imbalances between artifacts of different vintages, and inadequate organizational structures. These problems can be the direct or indirect result of climatic or geographic constraints, natural disasters, social and cultural demands for change, simple economic wants, military demands, varying resource positions, and other contingent factors.

The Green Revolution corroborates the claim that technological change is essentially a problem-solving activity. The actions undertaken by the various agents, specifically the government, were a clear response to a specific problem that was recognized, identified, and for which a specific plan of action was implemented. The problem of food shortage caused by the poor productivity of the land and other related issues were solved by the transfer, innovation, development, and diffusion of a new knowledge system. We saw that the political-economic and institutional factors helped to organize the policy framework, and ultimately the selection environment (metamarket). The technological change in the Green Revolution shows that it occurred as the outcome of a specific set of concrete actions and learning experiences of a wide array of human and social actors. It elucidates how the problems were defined and solutions were sought.

A major focusing device for the problem-solving heuristic
is provided by the functional problems of the technology

The driving force behind technological change is the limitation of the old technology and the constraints imposed on its functional limits. Increased productivity, improved efficiency, less cost, less human intervention, and other such improvement measures are normally achieved by improving on the existing technology. The starting point for thinking about technological change is naturally the functional failure or limitations of the old technology.

The inability to increase productivity beyond the subsistence level using the old tools and techniques was a direct correlate of their functional inadequacy. The constraints of the old system provided the problems that became the focus of new research and development activities. The solution was to come up with concrete measures to circumvent the productivity constraint. It started with the seed, irrigation technology, energy sources, and so on. The old seeds were replaced with HYVs. The problems associated with using these varieties further necessitated new approaches to agricultural practice. Constant renewal of seeds, increased usage of fertilizers and pesticides, water, and other inputs were mandated by the change of seeds.

The learning processes involved actions that necessitated substituting mechanical power for human and animal power, chemical fertilizers for manure, HYVs of seeds for traditional seeds, and irrigated water for rain water. The Green Revolution shows that human actions and learning experiences were responsible for the increase in cereal productivity rather than some *deus ex machina* from which these material benefits emanated.

An algorithm of decision rules involving the transformation
of newly derived and learned knowledge to solve problems
characterizes technological change

Technological change follows a clear heuristic in which the relevant structural components such as invention, innovation, development, and transfer interactively bring about the change. It involves the evolution of a technological algorithm of how newly derived knowledge for problem solving can be transformed into material outputs or into appropriate functional outcomes. The technological algorithm can be the means by which the complex knowledge of the experts can be simplified and translated for other "lower-level" agents, for example workers, extension agents, and farmers who actually carry out the desired actions to produce the "desired" output or service. This system can be equated to a decision-rule-making process by all the protagonists associated with the problem-solving community.

The Green Revolution reveals the evolution of a technological algorithm, demonstrating how new knowledge could be transformed into material output in the form of food. The algorithm provided the guidepost for translating complex knowledge of the experts into simplified information that the peasant-farmers could understand. They learned the new agricultural practice by trial and error, by observing how the new agriculture was practised. For them, the new agricultural technology became meaningful only after learning new ways of solving problems. The claim that technological change is knowledge change is manifested by the fact that their cognition of agricultural practice changed. This was not attributable to the introduction of a few new artifacts, but to a systematic change in the knowledge related to agriculture.

Technological change is molded and directed by social interests and forces

It has been shown that technological change is an evolutionary process and that it is possible to steer the course of development of technology in a "desirable" direction by influencing the "selection" environment. The technological selection process is characterized by instruction, understanding, experience of learning by doing, and finally by cognitive change. Technological change is construed as a selective-retention process that is adapted to a sequential process of variation and selection. Within the milieu of sociocultural evolution, adaptive learning and perception lead to the accumulation and change of technological knowledge. Technological knowledge is gained by trial and error, learning by doing, and learning by imitating.

Unlike natural selection where random processes determine the outcome, persuasion through propaganda and other incentive means and unanticipated and contingent factors decided the outcome of the technological selection process. The possibilities for "variation" are limited to the extent that market forces and the imperatives of public (mostly governmental) actions circumscribe the variation. This is because of the pre-established intention of changing the behavior of certain agents (for example, peasant-farmers in the Green Revolution). Thus, to a large extent, the public agencies, including market forces, can guide the selection process. Consequently, technological change is molded and directed by social forces and institutions such as governments, firms, households, markets, and so on. It is not an autonomous or deterministic process that is beyond the control of human agents.

The Green Revolution reinforces the claim that technological change is not an autonomous process that is out of societal control. It belies the theory of technological determinism. It was not the dictates of some internal logic of the technology that directed change and defined the technological trajectory. The problem-solving activity was entirely within social and institutional control. Public officials, research personnel, and aid agencies in India and abroad formulated the policies. It should be noted, however, that the power to formulate the policies and to create the selection environment was not equally distributed among all the actors. In constructing technological change, asymmetric power distribution is the norm rather the exception, because this process is social through and through. For example, the peasant-farmers did not have the choice to decide what technology they wanted.

Theories of scientific change may not be quite successful in explaining technological change

It was argued earlier that different cognitive and social forces, requiring different analytical means to study these forces, mold technology and science. This occurs despite the fact that technology and science relate interactively to benefit each other. The reasons why theories of scientific change are not the best explanatory tools to account for technological change are directly related to the differing social processes that organize scientific and technological knowledge production and dissemination. Technological change is driven by economic motives, in most circumstances, unless contingent factors like war or natural calamities and events might change the direction of new technology development. A form of action-oriented functional logic drives technological change. Scientific change, on the other hand, is driven by an explanation-oriented naturalistic logic. While technological knowledge is created for doing things, such as solving a practical problem (increasing food production), scientific knowledge is created for explaining and understanding nature and natural phenomena (how plants produce food through photosynthesis). Although these two can be conceptualized as problem-solving activities, the same analytical tools cannot represent them.[14]

It may be possible to find analogous situations in these phenomena and they may reflect common constraints, but may not reflect common causes. The appearances of entities that may resemble a paradigm are not enough to look for isomorphism between scientific change and technological change.

TECHNOLOGY POLICY IMPLICATIONS

The success of a model of technological change is its heuristic potential: its ability to represent all shapes and forms of this phenomenon. The universality of any model is its ability to show flexibility and adaptability to explain and understand various cases of technological change. Such a model should provide general ground rules for allowing how different component vectors of technological change can be represented, and how they interact to bring about the knowledge change. In this sense, the Green Revolution case study is only the beginning of an empirical quest. Despite its comprehensive nature, no claim is made that it is an exemplar. The previous checklist of Eleven Theses on technological change could be used as a general guide to develop a comprehensive model of technological change. The claim that technological change may be best understood as a problem-solving activity and that it may be best explained as a process of knowledge change is an early attempt at such a comprehensive theory.

The use-value of a good theory or model is its ability to explain and predict (certain aspects or features) successfully a phenomenon, natural or social. All correct explanations and predictions may lead to successful interventions in nature and society. Thus, the above characterization that technological change is a problem-solving activity and that it can be best explained as knowledge change has great technology policy implications. Governments or public agencies interested in effecting technological change can thus formulate the "right" technology policies. For example, governments and development agencies can formulate and execute the right policies to effectively transfer technologies from one country to another, from one region to another, and from universities and research establishments to industries, once these powerful social agents have learned how this process actually works. Once they have understood that technology is more than material artifacts and that technological change is essentially knowledge change, they may be able to create the appropriate environment to stimulate the technological selection process.

The arguments presented in this book have relevance to help formulate strategic technology policies in order to bring about technological change. For developing countries interested in improving the standard of living of their citizens through implementing effective technology transfer programs, the availability of the right models of technological change is crucial. The claim that technology is knowledge, and that technological change is knowledge change of the agents, is the foundation on which to develop such policies. The details of how such a policy framework can be developed may be gleaned from the various models analyzed in this book and from the Eleven Theses presented earlier. Also, for those interested in the intellectual ramifications of technological change regardless of its practical significance, the comprehensive analyses of technological change presented in this book should provide the agenda and inspiration for further study and research. What has been inside the "black box" is no mystery after all. Technological change is no more an abstraction than its obverse — society.

ENDNOTES

1. Barry Barnes, "The Science-Technology Relationship: A Model and a Query," *Social Studies of Science* 12 (1982): 166–172; and Edwin T. Layton, "Through the Looking Glass; or, News from Lake Mirror-Image," in *In Context: History and the History of Technology,* Stephen H. Cutcliffe and Robert C. Post, eds. (Bethlehem, PA: Lehigh University Press, 1989), 29–41.

2. Edwin T. Layton, "Technology as Knowledge," *Technology and Culture* 15, no. 1 (1974): 31–41; and Walter G. Vincenti, *What Engineers Know and How They Know It* (Baltimore: Johns Hopkins, 1990).

3. Layton, "Technology as Knowledge," and Eugene S. Ferguson, "The Mind's Eye: Nonverbal Thought in Technology," *Science* 197 (26 August 1977): 827–836.

4. Layton, "Technology as Knowledge."

5. Layton, "Technology as Knowledge"; Richard R. Nelson, "What Is Private and What Is Public About Technology?" *Science, Technology & Human Values* 14, no. 3 (1989): 229–241.

6. This claim may not hold any more as recent developments in genetics and molecular biology suggest. For example, multinational corporations that are investing large amounts of research money into human genome projects have been taking out patent rights on genetic information (such as certain gene sequences) so that they can develop medicines and gene therapy in the future for known or unknown diseases and human "deficiencies." Based on new trade and intellectual property regulations set up by the World Trade Organization, multinational biotechnology corporations like Monsanto, Merck, Cargill, and W. R. Grace are now actively funding projects in countries in the tropics for collecting the germ plasm, extracts, and essences of varieties of cereals, tubers, pulses, fungi, and medicinal plants for the sole purpose of developing proprietorial products and medicines for profits. Once the patents have been approved, the seeds, medicines, and other products developed by these corporations would become commodities. Those rooted forest peoples and rural dwellers who have been using this traditional knowledge would lose their right to freely plant and use them in the future. In 1992, W. R. Grace & Co. obtained a patent from the United States Patents and Trademarks Office for "neemix," a nontoxic (for humans) natural insecticide made from the seeds of the ancient *neem* tree that are found all over South Asia. Indian farmers have been using *neem*-based insecticides for hundreds of years. Another notorious case was the patent given to the medicinal property of turmeric to an American corporation. The point is that scientific knowledge may have already lost its status as certified public knowledge. For details of these two cases, see, respectively, A. Raver, "From the Ancient Neem Tree, a New Insecticide," *New York Times,* 5 June 1994, 49(N); and India Today, "Patents: Turmeric War," *India Today* (15 August 1996): 105.

7. Alois Huning, "Homo Mensura: Human Beings Are Their Technology—Technology Is Human," in *Research in Philosophy and Technology,* Vol. 8, Paul T. Durbin, ed. (Greenwich, CT: Jai Press, 1985), 9–16, points out that in the late nineteenth century Ernst Kapp in a book entitled *Grundlnien einer Philosophie der Technik* ["Foundations of a Philosophy of Technology"] (1877) used the term *organprojektion* or "organ projection" to refer to technology. Anthropological measure of technology as the extension of the human hand, in both the literal and metaphorical sense, is more than obvious. An excellent exposition of this idea can be found in Bruce Mazlish, *The Fourth Discontinuity: The Co-Evolution of Humans and Machines* (New Haven, CT: Yale University Press, 1993), who develops a coevolutionary theory of human and machine development.

8. Trevor Pinch, "The Social Construction of Technology: A Review," in *Technological Change: Methods and Themes in the History of Technology,* Robert Fox, ed. (Amsterdam: Harwood Academic Publishers, 1996), 17–35.

9. Thomas P. Hughes, "The Seamless Web: Technology, Science, Etcetera, Etcetera," *Social Studies of Science* 16 (1986): 281–292.

10. Works of Bruno Latour, *Science in Action: How to Follow Scientists and Engineers Through Society* (Cambridge, MA: Harvard University Press, 1987); Trevor Pinch, "Social Construction"; and Wiebe E. Bijker, *Of Bicycles, Bakelites, and Bulbs: Toward a Theory of Sociotechnical Change* (Cambridge, MA: MIT Press, 1995), attempt to incorporate the dynamic interplay of power in their explanations and models of technological change.

11. It can be shown that among Western industrialized nations, the government played

a key role, particularly since the Second World War, in promoting nuclear, aeronautics, and space and electronics technologies.

12. Chris de Bresson, "The Evolutionary Paradigm and the Economics of Technological Change," *Journal of Economic Issues* 21, no. 2 (1987): 751–762; and Joel Mokyr, "Evolution and Technological Change: A New Metaphor for Economic History," in *Technological Change: Methods and Themes in the History of Technology,* Robert Fox, ed. (Amsterdam: Harwood Academic Publishers, 1996), 63–83.

13. For example, for an interesting and informative discussion of the different meanings of an artifact within the theoretical frame of the sociology of scientific knowledge, see Donald MacKenzie, "How Do We Know the Properties of Artefacts? Applying the Sociology of Knowledge to Technology," in *Technological Change: Methods and Themes in the History of Technology,* Robert Fox, ed. (Amsterdam: Harwood Academic Publishers, 1996), 247–263.

14. A caveat is in order. The reformulations of Thomas Kuhn's model of scientific change for technological change by Edward Constant (chapter two) and Giovanni Dosi (chapter five) were able to explain the cases of technological change they analyzed quite successfully. However, they had to take extreme liberty in reformulating Kuhn's paradigm shift model to the selection of technological paradigms. The major difficulty was that unlike scientific change, technological change is evolutionary and continuous. Their success may be confined to only explaining a component or certain artifactual facet of airplanes (Constant) and microelectronics (Dosi).

Bibliography

Abramovitz, Moses. "Resource and Output Trends in the United States since 1870." In *The Economics of Technological Change: Selected Readings,* ed. Nathan Rosenberg. Harmondsworth, UK: Penguin Books, 1971, 320–343.

Adas, Michael. *Machines as the Measure of Men: Science, Technology, and Ideologies of Western Dominance.* Ithaca, NY: Cornell University Press, 1989.

Aitken, Hugh G. J. *The Continuous Wave: Technology and American Radio, 1900–1932.* Princeton, NJ: Princeton University Press, 1985.

Alvares, Claude A. *Decolonizing History: Technology and Culture in India, China and the West 1492 to the Present Day.* New York: Apex Press, and Goa, India: The Other India Press, 1991.

Amin, Samir. *Eurocentrism.* New York: Monthly Review Press, 1989.

———. *Accumulation on a World Scale: A Critique of the Theory of Underdevelopment.* New York: Monthly Review Press, 1974.

Anderson, Robert S., Paul R. Brass, Edwin Levy, and Barrie M. Morrison, eds. *Science, Politics, and the Agricultural Revolution in Asia.* Boulder, CO: Westview Press, 1982.

Atkinson, Anthony B., and Joseph E. Stiglitz. "A New View of Technological Change." *Economic Journal* 79 (1969): 573–578.

Barker, Christopher J. "Frogs and Farmers: The Green Revolution in India and Its Murky Past." In *Understanding Green Revolution: Agrarian Change and Development Planning in South Asia,* eds. Tim P. Bayliss-Smith and Sudhir Wanmali. Cambridge: Cambridge University Press, 1984, 37–53.

Barnes, Barry. "The Science-Technology Relationship: A Model and a Query." *Social Studies of Science* 12 (1982): 166–172.

Basalla, George. *The Evolution of Technology,* eds. Tim P. Bayliss-Smith and Sudhir Wanmali. Cambridge: Cambridge University Press, 1988.

———. *Understanding Green Revolution: Agrarian Change and Development Planning in South Asia.* Cambridge: Cambridge University Press, 1984.

Beck, Ulrich. *Risk Society: Towards a New Modernity.* London: Sage Publications, 1992.

Beer, John J. "The Historical Relations of Science and Technology: Introduction." *Technology and Culture* 6, no. 4 (1965): 547–552.

Bell, Daniel. *The Frontiers of Knowledge.* Garden City, NY: Doubleday, 1975.

Bettelheim, Charles. *India Independent.* New York: Monthly Review Press, 1968.

Biggs, Stephen D. "Informal R&D." *Ceres* 13, no. 4 (1986): 23–26.

Bijker, Wiebe E. *Of Bicycles, Bakelites, and Bulbs: Toward a Theory of Sociotechnical Change.* Cambridge, MA: MIT Press, 1995.

———. "The Social Construction of Fluorescent Lighting, or How an Artifact Was Invented in Its Diffusion Stage." In *Shaping Technology/Building Society: Studies in Sociotechnical Change,* eds. Wiebe E. Bijker and John Law. Cambridge, MA: MIT Press, 1992, 75–104.

———. "The Social Construction of Bakelite: Towards a Theory of Invention." In *The Social Construction of Technological Systems: New Directions in the Sociology and History of Technology,* eds. Wiebe E. Bijker, Thomas P. Hughes and Trevor Pinch. Cambridge, MA: MIT Press, 1987, 159–187.

Bijker, Wiebe E., and John Law. "General Introduction." In *Shaping Technology/ Building Society: Studies in Sociotechnical Change,* eds. Wiebe E. Bijker and John Law. Cambridge, MA: MIT Press, 1992, 1–16.

Bijker, Wiebe E., Thomas P. Hughes, and Trevor Pinch, eds. *The Social Construction of Technological Systems: New Directions in the Sociology and History of Technology.* Cambridge, MA: MIT Press, 1987.

Binswanger, Hans P. "Induced Technical Change: Evolution of a Thought." In *Induced Innovation: Technology, Institutions, and Development,* eds. Hans P. Binswanger and Vernon W. Ruttan. Baltimore: Johns Hopkins University Press, 1978, 13–43.

Binswanger, Hans P and Vernon W. Ruttan, eds. *Induced Innovation: Technology, Institutions and Development.* Baltimore: Johns Hopkins University Press, 1978.

Binswanger, Hans P. and Vernon W. Ruttan. "Introduction." In *Induced Innovation,* eds. Binswanger and Ruttan. Baltimore: Johns Hopkins University Press, 1978, 1–12.

Blaug, Mark. "Kuhn Versus Lakatos, or Paradigms Versus Research Programmes in the History of Economics." In *Paradigm and Revolutions,* ed. Gary Gutting. South Bend, IN: University of Notre Dame Press, 1980, 137–159.

Bloor, David. *Knowledge and Social Imagery.* London: Routledge and Keegan Paul, 1976.

Blyn, George. *Agricultural Trends in India, 1891–1947: Output, Availability and Productivity.* Philadelphia, PA: University of Pennsylvania Press, 1966.

Bogard, William. *The Bhopal Tragedy: Language, Logic, and Politics in the Production of a Hazard.* Boulder, CO: Westview Press, 1989.

Borlaug, Norman. *The Green Revolution, Peace and Humanity.* Mexico City: CIMMYT Report and translation series no. 3, 1972.

Brass, Paul R. "Institutional Transfers of Technology: The Land-Grant Model and the Agricultural University at Pantnagar." In *Science, Politics, and the Agricultural Revolution,* eds. Robert Anderson, et al., Boulder, CO: Westview Press, 1982, 103–163.

Bronfenbrenner, Martin. "The 'Structure of Revolutions' in Economic Thought." *History of Political Economy* 3, no. 1 (1971): 136–151.

Bruland, Tine. "Industrial Conflict as a Source of Technical Innovation: Three Cases." *Economy and Society* 11, (1982): 91–121.

Bryant, Lynwood. "The Development of the Diesel Engine." *Technology and Culture* 17, no. 3 (1976): 432–446.

Callon, Michel. "Society in the Making: The Study of Technology as a Tool for Sociological Analysis." In *Social Construction,* eds. Wiebe Bijker et al., Cambridge, MA: MIT Press, 1987, 83–106.

———. "The Sociology of an Actor-Network: The Case of the Electric Vehicle." In *Map-*

ping the Dynamics of Science and Technology, eds. Michel Callon, John Law and Arie Rip. London: Macmillan, 1986, 19–34.

Campbell, Donald T. "Evolutionary Epistemology." In *The Philosophy of Karl Popper,* ed. Paul Arthur Schlipp. La Salle, IL: Open Court, 1974, 413–463.

Chandler, Robert F. "The Scientific Basis for the Increased Yield Capacity of Rice and Wheat, and Its Present and Potential Impact on Food Production in the Developing Countries." In *Food, Population, and Employment: The Impact of the Green Revolution,* eds. Thomas T. Poleman and Donald K. Freebairn. New York: Praeger, 1973, 25–43.

Chengappa, Raj. "Agriculture: A Golden Revival." *India Today* (April 15, 1989): 78–83.

Clark, John, Ian McLaughlin, Howard Rose, and Robin King, *The Process of Technological Change: New Technology and Social Choice in the Workplace.* Cambridge: Cambridge University Press, 1988.

Clark, Norman. "Similarities and Differences Between Scientific and Technological Paradigms." *Futures* 19, no. 1 (1987): 26–42.

Coats, Alfred W. "Is There a 'Structure of Scientific Revolutions' in Economics?" *Kyklos* 22, no. 2 (1969): 289–298.

Collins, Harry M. "Stages in the Empirical Programme of Relativism." *Social Studies of Science* 11, no. 1 (1981): 3–10.

Constant, Edward W. "The Social Locus of Technological Practice: Community, System, or Organization." In *Social Construction,* eds. Wiebe Bijker, et al., Cambridge, MA: MIT Press, 1987, 223–242.

———. *The Origins of the Turbojet Revolution.* Baltimore: Johns Hopkins, 1980.

———. "A Model for Technological Change Applied to the Turbojet Revolution." *Technology and Culture* 14, no. 4 (1973): 553–572.

Coombs, Rod, Paolo Saviotti, and Vivien Walsh. *Economics and Technological Change.* Totowa, NJ: Rowman & Littlefield, 1987.

Copp, Newton, and Andrew Zanella. *Discovery, Innovation, and Risk: Case Studies in Science and Technology.* Cambridge, MA: MIT Press, 1993.

Dalrymple, Dana G. "Changes in Wheat Varieties and Yields in the United States, 1919–1984." *Agricultural History* 64, no. 4 (1988): 20–36.

———. *Development and Spread of High-Yielding Wheat Varieties in Developing Countries.* Washington, DC: United States Agency for International Development, 1986.

———. *Development and Spread of High-Yielding Rice Varieties in Developing Countries.* Washington, DC: United States Agency for International Development, 1986.

———. "The Development and Adoption of High-Yielding Varieties of Wheat and Rice in Developing Countries." *American Journal of Agricultural Economics* 67, no. 5 (1985): 1067–1073.

———. "The Adoption of High-Yielding Grain Varieties in Developing Nations." *Agricultural History* 53, no. 4 (1979): 704–726.

Dantwala, Mohanlal L. "From Stagnation to Growth: Relative Roles of Technology, Economic Policy and Agrarian Institutions." In *Technical Change in Asian Agriculture,* ed. Richard T. Shand. Canberra: Australian National University, 1973, 259–281.

Darwin, Charles. *On the Origin of Species,* A Facsimile of the First Edition. Cambridge, MA: Harvard University Press, 1964.

Dasgupta, Biplab. "India's Green Revolution." *Economic and Political Weekly* 12, nos. 6–8 (February 1977): 241–260.

David, Paul A. "Clio and the Economics of QWERTY." *American Economic Review, AEA Papers and Proceedings, Economic History* 75, no. 2 (1985): 332–337.

———. *Technical Choice, Innovation, and Economic Growth.* New York: Cambridge University Press, 1975.

De Bresson, Chris. "Breeding Innovation Clusters: A Source of Dynamic Development." *World Development* 17, no. 1 (1989): 1–16.

———. *Understanding Technological Change.* Montreal and New York: Black Rose Books, 1987.

———. "The Evolutionary Paradigm and the Economics of Technological Change." *Journal of Economic Issues* 21, no. 2 (1987): 751–762.

De Camp, L. Sprague. *The Ancient Engineers.* New York: Ballantine Books, 1974.

DeGregori, Thomas R. *A Theory of Technology: Continuity and Change in Human Development.* Ames, IA: Iowa State University Press, 1985.

Denison, Edward. *The Sources of Economic Growth in the U. S. and the Alternatives Before Us.* New York: Committee for Economic Development, 1962.

Dirlik, Arif. "Place-based Imagination: Globalism and the Politics of Place." *Review* 22, no. 2 (1999): 151–188.

Dosi, Giovanni. "Technological Paradigms and Technological Trajectories." *Research Policy* 11, no. 3 (1982): 147–162.

Dyson, Esther. *Release 2.0: A Design for Living in the Digital Age.* New York: Broadway Books, 1997.

Ellul, Jacques. *The Technological Society.* New York: Knopf, 1964.

Elster, Jon. *Explaining Technical Change: A Case Study in the Philosophy of Science.* Cambridge: Cambridge University Press, and Oslo: Universitetsforlaget, 1983.

———. *Logic and Society: Contradictions and Possible Worlds.* Chilchester, MA: John Wiley & Sons, 1978.

Escobar, Arturo. *Encountering Development: The Making and Unmaking of the Third World.* Princeton, NJ: Princeton University Press, 1995.

Ezell, Edward C. "Review of *The Origins of the Turbojet Revolution* by Constant, E.W." *American Historical Review* 87 (1982): 155.

Fabricant, Solomon. *Basic Facts on Productivity Change.* New York: National Bureau of Economic Research, 1959.

Felipe, Jesus. *Total Factor Productivity Growth in East Asia: A Critical Survey.* Manila, Philippines: Asian Development Bank, EDRC Report Series No. 65, 1997.

Feller, Irwin. "Review of *Patterns of Technological Innovation* by Sahal, D." *Science* (July 2, 1982): 47.

Ferguson, Adam. *Essays on the History of Civil Society.* New York: Garland Publishing, 1971.

Ferguson, Eugene S. "The Mind's Eye: Nonverbal Thought in Technology," *Science* 197, (August 26, 1977): 827–836.

Fox, Robert. "Introduction: Methods and Themes in the History of Technology." In *Technological Change: Methods and Themes in the History of Technology,* ed. Robert Fox. Amsterdam: Harwood Academic Publishers, 1996, 1–15.

Frank, Andre G. *Dependent Accumulation and Underdevelopment.* New York: Monthly Review Press, 1979.

Fransman, Martin. "Conceptualising Technical Change in the Third World in the 1980s: An Interpretive Survey." *Journal of Development Studies* 21, no. 4 (1985): 572–652.

Fuentes, Marta, and Andre G. Frank. "Ten Theses on Social Movements." *World Development* 17, no. 2 (1989): 179–191.

Fuller, Steve. "Being There With Thomas Kuhn: A Parable for Postmodern Times." *History and Theory* 31 (1992): 241–275.

Gamser, Matthew S. "Innovation, Technical Assistance, and Development: The Importance of Technology Users." *World Development* 16, no. 8 (1988): 711–720.

Giddens, Anthony. *Modernity and Self-Identity.* Stanford, CA: Stanford University Press, 1991.

———. *The Consequences of Modernity.* Stanford, CA: Stanford University Press, 1990.

Gieryn, Thomas F. "Boundary Work and the Demarcation of Science from Nonscience." *American Sociological Review* 48 (1983): 781–795.

Gilder, George. *Microcosm: The Quantum Revolution in Economics and Technology.* New York: Simon and Schuster, 1989.

Gilfillan, S. Colum. *The Sociology of Invention.* Cambridge, MA: MIT Press, 1970.

Gille, Bertrand. "Prolegomena to a History of Techniques." In *The History of Techniques, Volume I, Techniques and Civilizations,* ed. Bertrand Gille. New York: Gordon and Breach Science Publishers, 1986, 3–96.

Gills, Barry K, and Andre G. Frank. "The Cumulating of Accumulation: Thesis and Research for 5,000 Years of World System History." *Dialectical Anthropology* 15 (1990): 19–42.

Girifalco, Louis A. *Dynamics of Technological Change.* New York: Van Nostrand, 1991.

Glass, Colin J., and W. Johnson. "Metaphysics, MSRP, and Economics." *British Journal for the Philosophy of Science* 39 (1988): 313–329.

Grabowski, Richard. "The Theory of Induced Institutional Innovation: A Critique." *World Development* 16, no. 3 (1988): 385–394.

Griliches, Zvi. "Hybrid Corn and the Economics of Innovation." In *The Economics of Technological Change: Selected Readings,* ed. Nathan Rosenberg. Harmondsworth, UK: Penguin Books, 1971, 212–228.

Gunn, Richard. "Marxism and Philosophy: A Critique of Critical Realism." *Capital & Class,* no. 37 (1988): 87–116.

Hacking, Ian. *Representing and Intervening: Introductory Topics in the Philosophy of Science.* Cambridge: Cambridge University Press, 1983.

Hall, Bert. "Lynn White's *Medieval Technology and Social Change* After Thirty Years." In *Technological Change,* ed. Robert Fox. Amsterdam: Harwood Academic Publishers, 1996, 85–102.

Hall, A. Rupert. "Epilogue: The Rise of the West." In *A History of Technology,* Vol. III, eds. Charles J. Singer, E. J. Holmyard, A. Rupert Hall, and Trevor J. Williams. Oxford: Clarendon Press, 1957, 709–721.

Hanieski, John F. "The Airplane as an Economic Variable: Aspects of Technological Change in Aeronautics, 1903–1955." *Technology and Culture* 14, no. 4 (1973): 535–552.

Hård, Mikael. *Machines Are Frozen Spirit: The Scientification of Refrigeration and Brewing in the 19th Century—A Weberian Interpretation.* Boulder, CO: Westview Press, and Frankfurt: Campus Verlag, 1994.

Hayami, Yujiro. "Elements of Induced Innovation: A Historical Perspective for the Green Revolution." *Explorations in Economic History* 8 (1971): 445–472.

Hayami, Yujiro, and Vernon W. Ruttan. *Agricultural Development: An International Perspective.* Revised and expanded edition. Baltimore: Johns Hopkins, 1985.

Hayami, Yujiro, and Vernon W. Ruttan. *Agricultural Development: An International Perspective.* Baltimore: Johns Hopkins, 1971.

Headrick, Daniel H. *The Tools of Empire: Technology and European Imperialism in the Nineteenth Century.* New York: Oxford University Press, 1981.

Heidegger, Martin. *The Question Concerning Technology and Other Essays.* New York: Harper & Row, 1977.

Heilbroner, Robert L. "Do Machines Make History?" *Technology and Culture* 8 (1967): 335-345.

Hempel, Carl G. *Philosophy of Natural Sciences.* Englewood Cliffs, NJ: Prentice-Hall, 1966.

Hilton, Rodney H., and P. H. Sawyer. "Technical Determinism: The Stirrup and the Plough." *Past and Present* 24 (1963): 90–100.

Hirsh, Richard F. *Technology and Transformation in American Electric Utility Industry.* Cambridge and New York: Cambridge University Press, 1989.

Holt, Richard. "Medieval Technology and the Historians: The Evidence for the Mill." In *Technological Change,* ed. Robert Fox. Amsterdam: Harwood Academic Publishers, 1996, 103–121.

Hopper, David W. "Distortions of Agricultural Development Resulting from Government Prohibitions." In *Distortions in Agricultural Incentives,* ed. Theodore W. Schultz. Bloomington, IN: Indiana University Press, 1978, 69–78.

Hughes, Thomas P. "The Evolution of Large Technological Systems." In *Social Construction,* eds. Wiebe W. Bijker, et al., Cambridge, MA: MIT Press, 1987, 51–82.

———. "The Seamless Web: Technology, Science, Etcetera, Etcetera." *Social Studies of Science* 16 (1986): 281–292.

———. *The Networks of Power: Electrification in Western Society, 1880–1930.* Baltimore: Johns Hopkins, 1983.

———. "The Development Phase of Technological Change." *Technology and Culture* 17, no. 3 (1976): 423–431.

Hume, David. *Writings on Economics,* ed. and intro. by E. Rotwein. Madison, WI: University of Wisconsin Press, 1955.

Humphrey, John W., John P. Oleson, and Andrew N. Sherwood, eds. *Greek and Roman Technology: Annotated Translation of Greek and Latin Texts and Documents.* London and New York: Routledge, 1998.

Huning, Alois. "Homo Mensura: Human Beings Are Their Technology—Technology Is Human." In *Research in Philosophy and Technology,* Vol. 8, ed. Paul T. Durbin. Greenwich, CT: Jai Press, 1985, 9–16.

India Today. "Patents: Turmeric War." *India Today* (August 15, 1996): 105.

Kateb, George. "Technology and Philosophy," *Social Research* 64 (1997): 1224–1246.

Kendrik, John W. *Productivity Trends in the United States.* Princeton, NJ: Princeton University Press, 1961.

Kenney, Martin. "Schumpeterian Innovation and Entrepreneurs in Capitalism: A Case Study of the U.S. Biotechnology Industry." *Research Policy* 15 (1986): 21–31.

Kim, J. L., and Lawrence J. Lau. "The Sources of Growth of the East Asian Newly Industrialized Countries." *Journal of the Japanese and International Economies* 8 (1994): 235–271.

Krugman, Paul R. "The Myth of Asia's Miracle." *Foreign Affairs* 73, no. 6 (1994): 62–78.

Kuhn, Thomas S. *The Structure of Scientific Revolutions.* 2nd ed. Chicago: University of Chicago Press, 1970 (first edition 1962).

Lakatos, Imre. *The Methodology of Scientific Research Programmes.* Cambridge: Cambridge University Press, 1978.

Landes, David. *The Unbound Prometheus: Technological Change and Industrial Development in Western Europe.* Cambridge: Cambridge University Press, 1969.

Lash, Scott. "Reflexive Modernization: The Aesthetic Dimension." *Theory, Culture and Society* 10, no. 1 (1993): 1–24.

Latour, Bruno. *Aramis, or, The Love of Technology.* Cambridge, MA: Harvard University Press, 1996.

———. "*The Prince* for Machines as Well as for Machinations." In *Technology and Social Processes,* ed. Brian Elliott. Edinburgh: Edinburgh University Press, 1988, 20–43.

———. *Science in Action: How to Follow Scientists and Engineers Through Society.* Cambridge, MA: Harvard University Press, 1987.

Lau, Lawrence J. "Technical Progress, Capital Formation and Growth of Productivity." In *Competitiveness in International Food Markets,* eds. Maury E. Bredahl, Philip C. Abbott, and Michael R. Reed. Boulder, CO: Westview Press, 1994, 145–167.

Laudan, Larry. *Progress and Its Problems: Towards a Theory of Scientific Growth.* Berkeley: University of California Press, 1977.

Laudan, Rachel. "Cognitive Change in Technology and Science." In *The Nature of Technological Knowledge: Are Models of Scientific Change Relevant?* ed. Rachel Laudan. Dordrecht: D. Reidel, 1984, 83–104.

Lave, Lester B. *Technological Change: Its Conception and Measurement.* Englewood Cliffs, NJ: Prentice-Hall, 1966.

Law, John. "The Anatomy of a Socio-Technical Struggle: The Design of the TSR 2." In *Technology and Social Process,* ed. Brian Elliot. Edinburgh: Edinburgh University Press, 1988, 44–69.

———. "The Structure of Sociotechnical Engineering—A Review of the New Sociology of Technology." *Sociological Review* 35 (1987): 404–425.

———. "On the Social Explanation of Technical Change: The Case of the Portuguese Maritime Expansion." *Technology and Culture* 28, no. 2 (1987): 227–252.

———. "Technology and Heterogeneous Engineering: The Case of Portuguese Expansion." In *Social Construction of Technological Systems,* eds. Wiebe E. Bijker, et al., Cambridge, MA: MIT Press, 1987, 111–134.

Law, John, and Wiebe E. Bijker. "Postscript: Technology, Stability, and Social Theory." In *Shaping Technology,* eds. Wiebe E. Bijker and John Law. Cambridge, MA: MIT Press, 1992, 290–308.

Law, J., and M. Callon. "The Life and Death of an Aircraft: A Network Analysis of Technical Change." In *Shaping Technology,* eds. Wiebe E. Bijker and John Law. Cambridge, MA: MIT Press, 1992, 21–52.

Layton, Edwin T. "Through the Looking Glass; or, News from Lake Mirror Image." In *In Context: History and the History of Technology,* eds. Stephen H. Cutcliffe and Robert C. Post. Bethlehem, PA: Lehigh University Press, 1989, 29–41.

———. "Technology as Knowledge." *Technology and Culture* 15, no. 1 (1974): 31–41.

———. "Mirror-Image Twins: The Communities of Science and Technology in 19th-Century America." *Technology and Culture* 12 (1971): 562–580.

Leaf, Murray J. *Songs of Hope: The Green Revolution in a Punjabi Village.* New Brunswick, NJ: Rutgers University Press, 1984.

Lele, Uma J., and Arthur A. Goldsmith. "The Development of National Agricultural Research Capacity: India's Experience with the Rockefeller Foundation and Its Significance for Africa." *Economic Development and Cultural Change* 37, no. 2 (1989): 305–344.

Lewis, W. Arthur. "Economic Development with Unlimited Supplies of Labour." In *The Economics of Underdevelopment,* eds. A. Agarwala and S. Singh. Oxford: Oxford University Press, 1954, 400–449.

Lewis. W. David. "Review of *The Origins of the Turbojet Revolution* by Constant, E.W." *Technology and Culture* 23, no. 3 (1982): 512–516.

Lloyd, Geoffrey E. R. *Greek Science After Aristotle.* New York: Norton, 1973.

Lower, Milton. "The Concept of Technology within the Institutionalist Perspective." In *Evolutionary Economics,* Volume I, ed. Marc R. Tool. Armonk, NY: M. E. Sharpe, 1988, 197–226.

Lukacs, Georg. *History and Class Consciousness: Studies in Marxist Dialectics.* Cambridge, MA: MIT Press, 1971.

Lucas, Robert E. "On the Mechanisms of Economic Development." *Journal of Monetary Economics* 22, no. 1 (1988): 3–42.

Lyotard, Jean-Francois. *The Postmodern Condition: A Report on Knowledge.* Minneapolis, MN: University of Minnesota Press, 1984.

MacKenzie, Donald. "How Do We Know the Properties of Artefacts? Applying the Sociology of Knowledge to Technology." In *Technological Change,* ed. Robert Fox. Amsterdam: Harwood Academic Publishers, 1996, 247–263.

———. "The Missile Accuracy: A Case Study in the Social Processes of Technological Change." In *Social Construction of Technological Systems,* eds. Wiebe E. Bijker, et al., Cambridge, MA: MIT Press, 1987, 195–222.

———. "Marx and the Machine." *Technology and Culture* 25 (1984): 473–502.

McElvey, Maureen. *Evolutionary Innovations: The Business of Biotechnology.* Oxford: Oxford University Press, 1996.

McNeill, William. "How the West Won." *New York Review of Books* XLV, no 7 (April 23, 1998): 37–39.

Mamdani, Mahmood. *The Myth of Population Control: Family, Caste, and Class in an Indian Village.* New York: Monthly Review Press, 1972.

Mansfield, Edwin. *Technological Change.* New York: W. W. Norton, 1968.

Marx, Karl. *Capital: A Critique of Political Economy,* Volume I, tr. Ben Fowkes. New York: Vintage Books, 1977.

———. *The Poverty of Philosophy.* Moscow: Progress Publishers, 1955.

Marx, Karl, and Friedrich Engels. *Manifesto of the Communist Party.* Chicago: Encyclopaedia Britannica, 1952.

Mazlish, Bruce. *The Fourth Discontinuity: The Co-Evolution of Humans and Machines.* New Haven, CT: Yale University Press, 1993.

McCombie, John. "Rhetoric, Paradigms, and the Relevance of the Aggregate Production Function." Paper presented at the Tenth Malvern Political Economy Conference, August 1996.

McIntire, C. T., and Marvin Perry. "Toynbee's Achievements." In *Toynbee: Reappraisals,* eds. C. T. McIntire and Marvin Perry. Toronto: University of Toronto Press, 1989, 3–31.

Meadows, Donella, Dennis Meadows, Jorgen Randers, and William Behrens. *Limits to Growth: Confronting Global Collapse.* New York: Universe Books, 1972.

Meier, Gerald M., ed. *Leading Issues in Economic Development.* New York: Oxford University Press, 1976.

Miller, Richard W. *Analyzing Marx: Morality, Power, and History.* Princeton, NJ: Princeton University Press, 1984.

Misa, Thomas J. "Retrieving Sociotechnical Change from Technological Determinism." In *Does Technology Drive History: The Dilemma of Technological Determinism,* eds. Merritt Roe Smith and Leo Marx. Cambridge, MA: MIT Press, 1994, 115–141.

———. "Theories and Models of Technological Change: Parameters and Purposes." *Science, Technology & Human Values* 17 (1992): 3–12.

Mitcham, Carl. *Thinking Through Technology: The Path Between Engineering and Philosophy.* Chicago: University of Chicago Press, 1994.

Mohan, Rakesh, D. Jha, and Robert Evenson. "The Indian Agricultural Research System." *Economic and Political Weekly* 8, no. 13 (1973): A21–A26.

Mokyr, Joel. "Evolution and Technological Change: A New Metaphor for Economic History." In *Technological Change,* ed. Robert Fox. Amsterdam: Harwood Academic Publishers, 1996, 63–83.

Moore, Barrington. *Social Origins of Dictatorship and Democracy.* Boston: Beacon Press, 1966.

Moseman, Albert H. *Building Agricultural Research Systems in the Developing Nations.* New York: The Agricultural Development Council, 1970.

Mukherjee, P. K., and Brian Lockwood. "High Yielding Varieties Programme in India: An Assessment." In *Technical Change in Asian Agriculture,* ed. Richard T. Shand. Canberra: Australian National University Press, 1973, 51–79.

Mumford, Lewis. *The Myth of the Machine: Technics and Human Development.* New York: Harcourt Brace Jovanovich, 1966.

———. *Technics and Civilization.* New York: Harcourt Brace Jovanovich, 1963.

Myrdal, Gunnar. *Economic Theory and Underdeveloped Regions.* Bombay: Vora & Co., 1958.

Needham, Joseph. *The Grand Titration: Science and Society in East and West.* London: Allen & Unwin, 1969.

———. *Science and Civilization in China,* Vol. 1–7. Cambridge: Cambridge University Press, 1954–1988.

Nelson, Richard R. "Technical Change as Cultural Evolution." In *Learning and Technological Change,* ed. Ross Thomson. New York: St. Martin's Press, 1993, 9–23.

———. "What Is Private and What Is Public About Technology?" *Science, Technology & Human Values* 14, no. 3 (1989): 229–241.

———. "The Economics of Invention: A Survey of the Literature." *Journal of Business* 32, no. 2 (1959): 101–127.

Nelson, Richard R., and Sidney G. Winter. *An Evolutionary Theory of Economic Change.* Cambridge, MA: Harvard University Press, 1982.

———. "In Search of a Useful Theory of Innovation." *Research Policy* 6 (1977): 36–76.

Noble, David F. *The Religion of Technology: The Divinity of Man and the Spirit of Invention.* New York: Knopf, 1997.

———. *Forces of Production: A Social History of Automation.* New York: Knopf, 1984.

Novick, Peter. *That Noble Dream: The "Objectivity Question" and the American Histori-cal Profession.* Cambridge and New York: Cambridge University Press, 1988.

Ogburn, William F. *Social Change With Respect to Cultural and Original Nature.* New York: Dell Publishing Co., 1966.

————. *On Culture and Social Change: Selected Papers,* ed. O. D. Duncan. Chicago: University of Chicago Press, 1964.

————. "Technology and the Standard of Living." *American Journal of Sociology* 60, no. 4 (1955): 380–386.

Parayil, Govindan. "Transcending Technological Pessimism: Reflections on an Alternative Technological Order." Working Papers in the Social Sciences, No. 43, Division of Social Science, Hong Kong University of Science and Technology, 28 December 1998.

————. "The 'Revealing' and 'Concealing' of Technology." *Southeast Asian Journal of Social Science* 26, no. 1 (1998): 17–28.

————. "Practical Reflexivity as a Heuristic for Theorizing Technological Change." *Technology In Society* 19, no. 1 (1997): 161–175.

————. "Economics and Technological Change: An Evolutionary Epistemological Inquiry." *Knowledge and Policy* 7, no. 1 (1994): 79–91.

————. "Models of Technological Change: A Critical Review of Current Knowledge." *History and Technology* 10 (1993): 105–126.

————. "The Green Revolution in India: A Case Study of Technological Change." *Technology and Culture* 33, no. 4 (1992): 737–756.

————. "Yearley's *Science, Technology and Social Change:* Review." *Social Epistemology* 6, no. 1 (1992): 57–63.

————. "Yearley's *Science, Technology and Social Change:* Reply." *Social Epistemology* 6, no. 1 (1992): 73–75.

————. "Technological Knowledge and Technological Change." *Technology In Society* 13, no. 2 (1991): 289–304.

————. "Schumpeter on Invention, Innovation, and Technological Change." *Journal of the History of Economic Thought* 13 (1991): 78–89.

————. "Technological Change as a Problem-Solving Activity." *Technological Forecasting and Social Change* 40, no. 3 (1991): 235–248.

————. "Technology as Knowledge: An Empirical Affirmation." *Knowledge: Creation, Diffusion, Utilization* (now *Science Communication*) 13, no. 1 (1991): 36–48.

————. "Book Review: *The Process of Technological Change: New Technology and Social Choice in the Workplace* by John Clark, et al." *Science, Technology & Human Values* 15, no. 1 (1990): 124–125.

Parsons, Talcott. *The Social System.* New York: Free Press, 1964.

Pickstone, John V. "Bodies, Fields, and Factories: Technologies and Understanding in the Age of Revolutions." In *Technological Change,* ed. Robert Fox. Amsterdam: Harwood Academic Publishers, 1996, 51–61.

Picon, Antoine. "Towards a History of Technological Thought." In *Technological Change,* ed. Robert Fox. Amsterdam: Harwood Academic Publishers, 1996, 37–49.

Pinch, Trevor. "The Social Construction of Technology: A Review." In *Technological Change,* ed. Robert Fox. Amsterdam: Harwood Academic Publishers, 1996, 17–35.

————. "Understanding Technology: Some Possible Implications of Work in the Sociology of Science." In *Technology and Social Processes,* ed. Brian Elliott. Edinburgh: Edinburgh University Press, 1988, 70–83.

Pinch, Trevor, and Wiebe E. Bijker. "The Social Construction of Facts and Artifacts: Or How the Sociology of Science and the Sociology of Technology Might Benefit Each Other." In *Social Construction,* eds. Wiebe E. Bijker, et al., Cambridge, MA: MIT Press, 1987, 17–50.

Pitt, Joseph C. " 'Style' and Technology." *Technology in Society* 10 (1988): 447–456.

Polanyi, Michael. *Personal Knowledge: Towards a Post-Critical Philosophy.* Chicago: University of Chicago Press, 1958.

Poleman, Thomas T., and Donald K. Freebairn, eds. *Food, Population, and Employment: The Impact of the Green Revolution.* New York: Praeger, 1973.

Popper, Karl. *The Myth of the Framework: In Defence of Science and Rationality.* London: Routledge, 1994.

———. *The Logic of Scientific Discovery.* London: Routledge, 1992.

———. *The Poverty of Historicism.* London: Ark Paperbacks, 1986.

———. *Conjectures and Refutations: The Growth of Scientific Knowledge.* New York: Basic Books, 1963.

Prasad, C. *Elements of the Structure and Terminology of Agricultural Education in India.* Paris: Unesco Press, 1981.

Price, de Solla D. J. "Notes Towards A Philosophy of the Science/Technology Interaction." In *The Nature of Technological Knowledge: Are Models of Scientific Change Relevant?* ed. Rachel Laudan. Dordrecht: D. Reidel, 1984, 105–114.

———. "On the Historiographic Revolution in the History of Technology: Comments on the Papers by Multhauf, Ferguson, and Layton." *Technology and Culture* 15, no. 1 (1974): 42–48.

———. "Is Technology Historically Independent of Science? A Study in Statistical Historiography." *Technology and Culture* 6, no. 4 (1965): 553–568.

Raver, A. "From the Ancient Neem Tree, a New Insecticide." *New York Times,* Sunday, 5 June 1994, 49(N).

Read, Hadley. *Partners With India: Building Agricultural Universities.* Urbana-Champaign: University of Illinois Press, 1974.

Reynolds, Terry S. "Review of *Networks of Power: Electrification in Western Society, 1880-1930.*" *Technology and Culture* 25 (1984): 644–647.

Rogers, Everett M. *Diffusion of Innovations,* 3rd ed. New York: Free Press, 1983.

Romer, Paul. "Idea Gaps and Object Gaps in Economic Development." *Journal of Monetary Economics* 32 (1993): 543–573.

———."Endogenous Technological Change." *Journal of Political Economy* 98 (1990): S71–S102.

Rorty, Richard. *Objectivism, Relativism, and Truth.* Cambridge: Cambridge University Press, 1991.

———. *Contingency, Irony, and Solidarity.* Cambridge: Cambridge University Press, 1989.

———. *Consequences of Pragmatism: Essays, 1972–1980.* Minneapolis: University of Minnesota Press, 1982.

Rosenberg, Nathan. *Exploring the Blackbox: Technology, Economics, and History.* Cambridge: Cambridge University Press, 1994.

———. *Inside the Black Box: Technology and Economics.* Cambridge: Cambridge University Press, 1982.

———. *Perspectives on Technology.* Cambridge and New York: Cambridge University Press, 1976.

————, ed. *The Economics of Technological Change: Selected Readings*. Harmondsworth, UK: Penguin Books, 1971.

————. "The Direction of Technological Change: Mechanisms and Focusing Devices." *Economic Development and Cultural Change* 18, no. 1 (1969): 1–24.

Rosenberg, Nathan, and Walter G. Vincenti. *The Britannia Bridge: The Generation and Diffusion of Technological Knowledge*. Cambridge, MA: MIT Press, 1978.

Rostow, Walt W. *The Stages of Economic Growth: A Non-Communist Manifesto,* 3rd ed. Cambridge and New York: Cambridge University Press, 1990.

Ruttan, Vernon W. "Induced Institutional Change." In *Induced Innovation,* eds. Binswanger and Ruttan. Baltimore: Johns Hopkins, 1978, 327–357.

————. "The International Agricultural Research Institute as a Source of Agricultural Development." *Agricultural Administration* 5, no. 4 (1978): 293–308.

————. "Technical and Institutional Transfer in Agricultural Development." *Research Policy* 4 (1975): 350–378.

————. "Usher and Schumpeter on Invention, Innovation and Technological Change." In *Economics of Technological Change,* ed. Rosenberg. Harmondsworth, UK: Penguin Books, 1971, 73–85.

Ruttan, Vernon W., and Hans Binswanger. "Induced Innovation and the Green Revolution." In *Induced Innovation,* eds. Binswanger and Ruttan. Baltimore: Johns Hopkins, 1978, 358–408.

Ruttan, Vernon W., and Yujiro Hayami. "Technology Transfer and Agriculture Development." *Technology and Culture* 14, no. 2 (1973): 119–151.

Sahal, Devendra. "Technological Guideposts and Innovation Avenues." *Research Policy* 14 (1985): 61–82.

————. *Patterns of Technological Innovation*. Reading, MA: Addison-Wesley Publishing, 1981.

————. "Alternative Conceptions of Technology." *Research Policy* 10 (1981): 2–24.

Samuelson, Paul. "Schumpeter as a Teacher and Economic Theorist." *Review of Economics and Statistics* 33 (1951): 98–103.

Saviotti, Paolo P. "Systems Theory and Technological Change." *Futures* 18, no. 6 (1983): 773–786.

Schmookler, Jacob. *Inventions and Economic Growth*. Cambridge, MA: Harvard University Press, 1966.

————. "The Changing Efficiency of the American Economy, 1869–1938." *Review of Economics and Statistics* 34 (1952): 214–231.

Schot, Johan W. "Constructive Technology Assessment and Technology Dynamics: The Case of Clean Technologies." *Science, Technology & Human Values* 17, no. 1 (1992): 36–56.

Schultz, Theodore W. "Reflections on Agricultural Production, Output and Supply," *Journal of Farm Economics* 38 (1956): 748–762.

Schumpeter, Joseph. "The Analysis of Economic Change." In *Essays of J. A. Schumpeter.* ed. R.V. Clemence. Cambridge, MA: Addison-Wesley, 1951, 134–142.

————. *Capitalism, Socialism, and Democracy*. New York: Harper & Row, 1950.

————. *Business Cycles,* 2 vols. New York: McGraw-Hill Press, 1939.

————. *The Theory of Economic Development*. Cambridge, MA: Harvard University Press, 1934.

Schweitzer, Paul R. "Usher and Schumpeter on Invention, Innovation and Technological Change: Comment." *Quarterly Journal of Economics* 75, no. 1 (1961): 152–156.

Sen, S. R. *Growth and Instability in Indian Agriculture,* Agriculture Situation in India XXI. New Delhi: Ministry of Food and Agriculture, 1967.

Shaikh, Anwar. "Laws of Production and Laws of Algebra: The Humbug Production Function." *Review of Economics and Statistics* 56 (1974): 115–120.

Shand, Richard T., ed. *Technical Change in Asian Agriculture.* Canberra: Australian National University Press, 1973.

Shaw, William H. " 'The Hand-mill gives you the Feudal Lord': Marx's Technological Determinism." *History and Theory* 18 (1979): 155–166.

Simon, Herbert A. "On Parsimonious Explanations of Production Relations." *Scandinavian Journal of Economics* 89 (1979): 459–474.

Singh, D. P. "Agricultural Universities and Transfer of Technology in India: The Importance of Management." In *Science, Politics,* eds. Robert Anderson et al., 165–178.

Sismondo, Sergio. "Some Social Constructions." *Social Studies of Science* 23 (1993): 515–553.

Smith, Adam. *An Inquiry into the Nature and Causes of the Wealth of Nations.* Chicago: University of Chicago Press, 1976.

Smith, John K. "Thinking about Technological Change: Linear and Evolutionary Models." In *Learning and Technological Change,* ed. Ross Thomson. New York: St. Martin's Press, 1993, 65–78.

Solo, Carolyn S. "Innovation in the Capitalist Process: A Critique of the Schumpeterian Theory." *Quarterly Journal of Economics* 65, no. 3 (1951): 417–428.

Solow, Robert. "Investments and Technical Progress." In *Mathematical Methods in the Social Sciences, 1959,* eds. Kenneth J. Arrow, Samuel Karlin, and Patrick Suppes. Stanford, CA: Stanford University Press, 1960, 89–104.

———. "Technical Change and the Aggregate Production Function." *Review of Economics and Statistics* 39 (1957): 312–320.

———. "A Contribution to the Theory of Economic Growth." *Quarterly Journal of Economics* 81 (1956): 65–94.

Spengler, Oswald. *The Decline of the West,* 2 volumes. New York: Knopf, 1992.

Spier, Robert F. G. *From the Hand of Man: Primitive and Preindustrial Technologies.* Boston: Houghton Mifflin, 1970.

Staudenmaier, John M. *Technology's Storyteller: Reweaving the Human Fabric.* Cambridge, MA: MIT Press, 1985.

Stoneman, Paul. *The Economic Analysis of Technology Policy.* Oxford: Clarendon Press, 1987.

Strassmann, W. Paul. "Creative Destruction and Partial Obsolescence in American Economic Development." *Journal of Economic History* 14, no. 3 (1959): 335–349.

Streeten, Paul. "The Use and Abuse of Models in Development Planning." In *Leading Issues,* ed. Gerald M. Meier. New York: Oxford University Press, 1976, 875–889.

Subramaniam, C. *The New Strategy in Indian Agriculture.* New Delhi: Vikas Publishing House, 1979.

Sud, Surinder. "State of Agriculture: India Poised for Take-Off." *Times of India* (19 January 1989), 7(N).

Susskind, Charles. *Understanding Technology.* Baltimore: Johns Hopkins, 1973.

Taviss, Irene. "Review of *Supplement to the Sociology of Invention* by Gilfillan, S.G." *Technology and Culture* 15, no. 1 (1974): 136–138.

Thomas, Donald E. *Diesel: Technology and Society in Industrial Germany.* Tuscaloosa, AL: University of Alabama Press, 1987.

Toulmin, Stephen. *Human Understanding,* Vol. I. Princeton, NJ: Princeton University Press, 1971.

Toynbee, Arnold. *A Study of History.* Revised and abridged. London: Oxford University Press, 1962.

Usher, Abbott Payson. "Technical Change and Capital Formation." In *Economics of Technological Change,* ed. Nathan Rosenberg. Harmondsworth, UK: Penguin Books, 1971, 43–72.

———. *A History of Mechanical Inventions.* Cambridge, MA: Harvard University Press, 1954.

Van den Belt, Henk, and Arie Rip. "The Nelson-Winter-Dosi Model and Synthetic Dye Chemistry." In *Social Construction,* eds. Wiebe E. Bijker, et al., Cambridge, MA: MIT Press, 1987, 135–158.

Van Weigel, B. "The Basic Needs Approach: Overcoming the Poverty of Homo Oeconomics." *World Development* 14, no. 12 (1986): 1423–1434.

Vincenti, Walter G. *What Engineers Know and How They Know It.* Baltimore: John Hopkins, 1990.

———. "Technological Knowledge without Science: The Invention of Flush Riveting in American Airplanes, ca. 1930–ca. 1950." *Technology and Culture* 25 (1984): 540–576.

Von Hippel, Eric. *The Sources of Innovation.* New York: Oxford University Press, 1988.

Wallerstein, Immanuel. *The Modern World System I: Capitalist Agriculture and the Origins of the European World-Economy.* San Diego: Academic Press, 1974.

Walters, Alan A. "Production and Cost Functions: An Econometric Survey." *Econometrica* 31 (1963): 1–66.

Weber, Max. *The Protestant Ethic and the Spirit of Capitalism,* tr. Talcott Parsons. London: Harper Collins Academic, 1991.

———. *General Economic History.* New Brunswick, NJ: Transaction Books, 1981.

Westrum, Ron. "The Social Construction of Technological Systems. Review of Bijker, et al. (eds.), *The Social Construction of Technological Systems.*" *Social Studies of Science* 19 (1989): 189–191.

Whewell, William. *The Philosophy of the Inductive Sciences,* Vol. 1, "Of Ideas in General." New York and London: Johnson Reprint Corporation, 1967.

———. *Philosophy of the Inductive Sciences,* Vol. 2, "Of Knowledge." New York and London: Johnson Reprint Corporation, 1967.

White, Hayden. *Tropics of Discourse: Essays in Cultural Criticism.* Baltimore: Johns Hopkins, 1985.

White, Lynn, Jr. "The Historical Roots of Our Ecological Crisis." *Science* 155 (10 March 1967): 1203–1207.

———. *Medieval Technology and Social Change.* Oxford: Oxford University Press, 1962.

Winner, Langdon. "Technology Today: Utopia or Dystopia." *Social Research* 64 (1997): 989–1017.

———. "Upon Opening the Black Box and Finding It Empty: Social Constructivism and the Philosophy of Technology." *Science, Technology & Human Values* 18 (1993): 362–378.

———. *The Whale and the Reactor: A Search for Limits in an Age of High Technology.* Chicago: University of Chicago Press, 1986.

————. *Autonomous Technology: Technics-out-of-Control as a Theme in Political Thought.* Cambridge, MA: MIT Press, 1977.

Wise, George. "Science and Technology." *OSIRIS,* 2nd series, 1 (1985): 229–246.

Wolfson, Robert J. "The Economic Dynamics of Schumpeter." *Economic Development and Cultural Change* 7 (1958/59): 31–54.

Woolgar, Steve. "The Turn to Technology in Social Studies of Science." *Science, Technology & Human Values* 16 (1991): 20–50.

Yearley, Steven. *Science, Technology, and Social Change.* London: Unwin Hyman, 1988.

Index

About the Author

An engineer and social scientist, Govindan Parayil has been on the faculty of the Division of Social Science at the Hong Kong University of Science and Technology since 1994. He teaches and researches on topics in science, technology, and society (STS), and environmental studies. He earned a Ph.D. in Science and Technology Studies from Virginia Tech. He has been a Rockefeller Foundation post-doctoral fellow in the Department of Science and Technology Studies at Cornell University (1993–1994). He also has held visiting positions at Illinois Institute of Technology (1991–1992) and Virginia Tech (1990–1991).

Parayil has published widely in the field, with articles appeared in *Technology and Culture, Technological Forecasting and Social Change, Technology in Society, Third World Quarterly, Journal of the History of Economic Thought, Global Environmental Change, Journal of Applied Philosophy,* and *Science Studies.* He is the editor of *Kerala's Development Experience: Reflections on Sustainability and Replicability.* Parayil is currently at work on a book on the philosophy and social studies of technology, *Modern Technological Order and Its Discontents.*

DATE DUE

MAY 0 4 2000			
3 2000			
MAY 2 2001			
MAY 1 2 2001			
MAY 0 1 2001			
AUG 1 1 2001			
AUG 1 0 2001			
MAY 1 2002			
JUN 0 3 2002			
NOV 2 9 2011			

BOWLING GREEN STATE UNIVERSITY DISCARDED LIBRARY

GAYLORD

PRINTED IN U.S.A.

SCI T 14.5 .P37 1999

Parayil, Govindan, 1955-

Conceptualizing
 technological change